SMP interact

C1

Teacher's guide to Book C1

CAMBRIDGE
UNIVERSITY PRESS

PUBLISHED BY THE PRESS SYNDICATE OF THE UNIVERSITY OF CAMBRIDGE
The Pitt Building, Trumpington Street, Cambridge, United Kingdom

CAMBRIDGE UNIVERSITY PRESS
The Edinburgh Building, Cambridge CB2 2RU, UK
40 West 20th Street, New York, NY 10011–4211, USA
10 Stamford Road, Oakleigh, VIC 3166, Australia
Ruiz de Alarcón 13, 28014 Madrid, Spain
Dock House, The Waterfront, Cape Town 8001, South Africa

http://www.cambridge.org

First published 2001
Reprinted 2001

Printed in the United Kingdom at the University Press, Cambridge

Typeface Minion   *System* QuarkXPress®

A catalogue record for this book is available from the British Library

ISBN 0 521 79863 9 paperback

Typesetting and technical illustrations by The School Mathematics Project
Other illustrations by Robert Calow and Steve Lach at Eikon Illustration
Photograph by Graham Portlock
Cover image © Tony Stone Images/Daryl Torckler
Cover design by Angela Ashton

Contents

The following people contributed to the writing of the SMP Interact key stage 3 materials.

Ben Alldred	Ian Edney	John Ling	Susan Shilton
Juliette Baldwin	Steve Feller	Carole Martin	Caroline Starkey
Simon Baxter	Rose Flower	Peter Moody	Liz Stewart
Gill Beeney	John Gardiner	Lorna Mulhern	Pam Turner
Roger Beeney	Bob Hartman	Mary Pardoe	Biff Vernon
Roger Bentote	Spencer Instone	Peter Ransom	Jo Waddingham
Sue Briggs	Liz Jackson	Paul Scruton	Nigel Webb
David Cassell	Pamela Leon	Richard Sharpe	Heather West

Others, too numerous to mention individually, gave valuable advice, particularly by commenting on and trialling draft materials.

Editorial team:	David Cassell	Project Administrator:	Ann White
	Spencer Instone	Design:	Melanie Bull
	John Ling		Tiffany Passmore
	Mary Pardoe		Martin Smith
	Paul Scruton	Project support:	Carol Cole
	Susan Shilton		Pam Keetch
			Nicky Lake
			Jane Seaton
			Cathy Syred

Special thanks go to Colin Goldsmith.

Introduction

What is distinctive about *SMP Interact*?

SMP Interact sets out to help teachers use a variety of teaching approaches in order to stimulate pupils and foster their understanding and enjoyment of mathematics.

A central place is given to discussion and other interactive work. Through discussion with the whole class you can find out about pupils' prior understanding when beginning a topic, can check on their progress and can draw ideas together as work comes to an end. Working interactively on some topics in small groups gives pupils, including the less confident, a chance to clarify and justify their own ideas and to build on, or raise objections to, suggestions put forward by others.

Questions that promote effective discussion and activities well suited to group work occur throughout the material.

SMP Interact has benefited from extensive and successful trialling in a variety of schools. The practical suggestions contained in the teacher's guides are based on teachers' experiences, often expressed in their own words.

Who are Books C1 to 3 for?

Books C1 to 3 follow on from Books 1 and N and cover national curriculum levels up to 7. (Level 8 can be covered from Book H1, the first of the higher tier KS4 books.)

How are the pupils' books intended to be used?

The pupils' books are a resource which can and should be used flexibly. They are not for pupils to work through individually at their own pace. Many of the activities are designed for class or group discussion.

Activities intended to be led by the teacher are shown by a solid strip at the edge of the pupil's page, and a corresponding strip in the margin of the teacher's guide, where they are fully described.

A broken strip at the edge of the page shows an activity or question in the pupil's book that is likely to need teacher intervention and support.

Where the writers have a particular way of working in mind, this is stated (for example, 'for two or more people').

Where there is no indication otherwise, the material is suitable for pupils working on their own.

Starred questions (for example, *C7) are more challenging.

What use is made of software?

Points at which software (on a computer or a graphic calculator) can be used to provide effective support for the work are indicated by these symbols, referring to a spreadsheet, graph plotter or dynamic geometry package respectively. Other suggestions for software support can be found on the SMP's website: www.smpmaths.org.uk

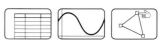

How is the attainment of pupils assessed?

The interactive class sessions provide much feedback to the teacher about pupils' levels of understanding.

Each unit of work begins with a statement of the key learning objectives and finishes with questions for self-assessment ('What progress have you made?') The latter can be incorporated into a running record of progress.

Revision questions are included in periodic reviews in the pupil's book.

A pack of assessment materials for Books T1, S1 and C1 contains photocopiable masters providing a short assessment for most of the units. Enclosed with the pack is a CD-ROM holding the assessment materials in question bank form so you can compile and edit tests on screen to meet your school's needs. Details of the pack are on the SMP's website.

What will pupils do for homework?

The practice booklets may be used for homework.

Often a homework can consist of preparatory or follow-up work to an activity in the main pupil's book.

Answers to questions on resource sheets

For reasons of economy, where pupils have to write their responses on a resource sheet the answers are not always shown in this guide. For convenience in marking you could put the correct responses on a spare copy of each sheet and add it to a file for future use.

General guidance on teaching approaches

Getting everyone involved

When you are introducing a new idea or extending an already familiar topic, it is important to get as many pupils as possible actively engaged.

Posing key questions
A powerful technique for achieving this is to pose one or two key questions, perhaps in the form of a novel problem to be solved. Ask pupils, working in pairs or small groups, to think about the question and try to come up with an answer.

When everyone has had time to work seriously at the problem (have a further question ready for the faster ones), you can then ask for answers, without at this stage revealing whether they are right or wrong (so that pupils have to keep thinking!). You could ask pupils to comment on each other's answers.

Open tasks
Open tasks and questions are often good for getting pupils to think, or thinking more deeply. For example, 'Working in groups of three or four, make up three questions which can be solved using the technique we have just been learning. Try to make your questions as varied as possible.'

Questioning skills

Questioning with the whole class
If your questions to the class are always closed, and you reward the first correct response you get, then you have no way of telling whether other pupils knew the correct answer or whether they had thought about the question at all. It is better to try to get as many pupils as possible to engage with the question, so do not at first say whether an answer is right or wrong. You could ask a pupil how they got their answer, or you could ask a second pupil how they think the first one got their answer.

Working in groups

Types of group work
Group work may be small scale or large scale. In small scale group work, pupils are asked to work in pairs or small groups for a short while, perhaps to come up with a solution to a novel type of problem before their suggestions are compared. In large scale group work, pupils carry out in groups a substantial task such as planning a statistical inquiry or designing a poster to get over the essential idea of the topic they have just been studying.

Organising the groups
Group size is important. Groups of more than four or five can lead to some pupils making little or no contribution.

For some activities, you may want pupils to work unassisted. But for many, your own contribution will be vital. Then it is generally more effective if, once you are sure that every group has got started, you work intensively with each group in turn.

After the group work One way to help pupils feel that they have worked effectively is to get them to report their findings to the whole class. This may be done in a number of different ways. One pupil from each group could report back. Or you could question each group in turn. Or each group could make a poster showing their results.

Managing discussion

Discussion, whether in a whole-class or group setting, has a vital role to play in developing pupils' understanding. It is most fruitful in an atmosphere where pupils know their contributions are valued and are not always judged in terms of immediate correctness. It needs careful management for it to be effective and teachers are often worried that it will get out of hand. Here are a few common worries, and ways of dealing with them.

What if ... '... the group is not used to discussion?'

- Allow time for pupils to work first on the problem individually or in small groups, then they will all have ideas to contribute.

'... everyone tries to talk at once?'

- Set clear rules. For example, pupils raise their hands and you write their name on the board before they can speak.

'... a few pupils dominate whole-class discussion?'

- Precede any class discussion with small-group discussion and nominate the pupils who will feed back to the class.

'... one pupil reaches the end point of a discussion immediately?'

- Tell them that the rest of the group need to be convinced and ask the pupil to convince the rest of the group.

- Accept the suggestion and ask the rest of the group to comment on it.

Action and result puzzles (p 4)

In each puzzle, the action cards show operations to be performed on a starting number and the result cards show the results. Pupils match up the results with the actions.

The puzzles provide number practice and an opportunity to apply some logical thinking. They may reveal misconceptions about number.

Essential	**Optional**
Puzzles on sheets 108 and 109 Scissors	Sheet 110 (blank cards) Puzzles on sheet 107 (easier) OHP transparencies of some sheets, cut into puzzle cards

'I put copies of the games into labelled boxes. I explained what each game was about and allowed them to choose their own games.'

A selection from the puzzles should be used (listed below roughly in order of difficulty).

Sheet 108   16.5 puzzle ($+$, $-$, $\times$ and $\div$, decimals and negative numbers)

           s puzzle (the number s has to be found)

           12 puzzle (a, b, c and d are numbers to be found)

Sheet 109   8 puzzle ($\times$ and $\div$, including fractions, simplifying fractions)

           20 puzzle (a, b, c and d are numbers to be found)

Pupils who need a gentler start could do the following.

Sheet 107   36 puzzle ($+$, $-$, $\times$ and $\div$, simple decimals and negative numbers)

           q puzzle ($+$ and $-$, two-digit numbers, q has to be found)

           h puzzle ($+$, $-$, $\times$ and $\div$, two-digit numbers, h has to be found)

◊ This has worked well with pupils sitting in pairs at tables of four. When each pair had matched the cards, all four pupils discussed what they had done. An aim is to encourage mental number work. However, pupils may want to do some calculations and demonstrate things to their group using pencil and paper. A calculator should not be used.

'I copied the cards on to pieces of acetate which could be moved about on the OHP. Pupils went to the OHP to show how the cards matched up.'

◊ Puzzles that pupils find easy can be done without cutting out the cards: they simply key each action card to its result card by marking both with the same letter. However, something may be learnt from moving cards around to try ideas out before reaching a final pairing, and some puzzles are almost impossible unless they are done this way.

◊ Solutions can be recorded by
- keying cards to one another with letters as described above
- sticking pairs of cards on sheets or in exercise books
- writing appropriate statements

◊ After pupils have solved some puzzles, they can make up some of their own (using the blank cards) to try on a partner. This may tell you something about the limits of the mathematics they feel confident with. Most should be able to make up puzzles of the *s* and 12 types.

◊ **Dividing by numbers between 0 and 1**

Some pupils were successful at reaching understanding by themselves and went on to explain division by fractions to other pupils. They thought about division as 'how many there are in' and reasoned, for example, from 'there are four 2s in 8' to 'there are sixteen halves in 8'.

Sheet 108

16.5 puzzle

Action	Result
+ 3.5	20
÷ 10	1.65
− 17	⁻0.5
× ⁻2	⁻33
÷ 0.5	33
+ ⁻3.5	13
− 6.05	10.45
× 0.1	1.65

s puzzle: $s = 4.5$

12 puzzle: $a = 0.5$, $b = 4$ and $c = 2$ (or $b = 2$ and $c = 4$), $d = 10$

Sheet 109

8 puzzle

Action	Result	Action	Result
÷ $\frac{1}{2}$	16	× 2	16
÷ 4	2	÷ 12	$\frac{2}{3}$
× $\frac{1}{8}$	1	× $\frac{1}{10}$	$\frac{4}{5}$
÷ $\frac{1}{4}$	32	÷ $\frac{1}{8}$	64
÷ 5	$1\frac{3}{5}$	× 4	32
× 10	80	÷ 8	1
÷ 2	4	÷ 10	$\frac{4}{5}$
× $\frac{1}{2}$	4	× $\frac{1}{4}$	2

20 puzzle: $a = \frac{1}{2}$, $b = ⁻20$, $c = 0.1$, $d = 40$

Sheet 107 (gentler start)

36 puzzle

Action	Result
÷ 9	4
− 40	⁻4
× 3	108
+ 27	63
÷ 8	4.5
× 1.5	54
+ ⁻50	⁻14
÷ 24	1.5

q puzzle: $q = 27$

h puzzle: $h = 9$

② Chocolate (p 4)

This is a problem-solving activity. It gives pupils an opportunity to compare fractions, but is presented in such a way that they have to think out an approach for themselves.

Essential	Optional
Bars or blocks of something which can be divided up and shared out equally	Bars of chocolate (of a kind not already subdivided into portions)

◊ The problems are all variations of this basic idea:
  • A number of tables are set out, each with some chocolate bars.
  • A group of pupils are asked, one by one, to choose a table to sit at.
  • When everyone has sat at a table, the bars on each table are shared equally between those at that table.

The problem for the pupils is to decide which table to sit at in the hope of getting the most chocolate at the shareout.

Getting started

◊ It is best to start with a fairly simple situation, for example three tables with 1, 2 and 3 bars.

Choose a group of pupils to take part, say eight, and explain the problem. Ask them one by one to choose their table (they cannot change their minds later).

◊ As pupils choose where to sit, involve the whole class and ask questions such as:
  • Where would you sit? Why?
  • How much chocolate would each person get at this table if no one else sits here?
  • Is it best to be the first to choose, the last, or doesn't it matter?

◊ Once the last pupil has chosen, ask pupils to decide who gets the most chocolate and to justify their answer. Pupils could consider this in small groups and then give their explanations to the whole class.

Explanations could involve fractions, percentages, decimals, ratios, or they may be idiosyncratic (for example, giving each bar a particular weight and dividing).

Variations

◊ Obviously the number of tables and/or bars can be varied.

You could also tell pupils that you will decide beforehand how many pupils will sit down but they will not know this until you stop them and they then share out the chocolate.

Follow-up work

◊ Pupils could work in pairs or small groups on particular problems, such as

  • Is it always the best strategy for the first person to sit at the table with most chocolate?

For example, suppose there are three pupils and three tables with 1, 2 and 3 bars.

Suppose the first pupil goes to the table with 3 bars.
The second pupil's best choice appears to be the table with 2 bars.
The third pupil's best choice is the table with 3 bars.

The second pupil has done best. So the first pupil would have done better to have gone to this table.

Chance

This unit introduces probability through games of chance. Probabilities are based on equally likely outcomes and the work here includes equivalent fractions.

Essential	**Optional**
Dice, counters Sheets 111 to 115	OHP transparencies of sheets 114 and 115
Practice booklet pages 3 to 5	

A **Chance or skill** (p 5)

Dice and counters, sheets 111 to 113 (game boards)

◊ Before discussing and playing the games, you could get pupils talking about chance, e.g. the National Lottery. People often have peculiar ideas about chance. For example, would they write on a National Lottery ticket the same combination as the one which won last week? If not, why not?

You could ask pupils to think about games they know and to discuss the elements of chance and skill in them.

◊ Before playing the games, ask pupils to think about each game in advance and to try to decide from its rules whether it is a game of pure chance, a game of skill, or a mixture.

Some games of skill give an advantage to the first player. Who goes first is usually decided by a process of chance.

◊ You could split the class into pairs or small groups, with each group playing one of the games and reporting on it.

◊ 'Fours' is a game of skill. 'Line of three' is a mixture of chance and skill. 'Jumping the line' appears to involve skill, because you have to decide which counters to move and it looks as if you can get 'nearer' to winning. But it is a game of pure chance. At any stage there is only one number which will enable the player to win. If any other number comes up, whatever the player does leaves the opponent in essentially the same position.

B Fair or unfair? (p 6)

> Dice, counters, sheets 114 and 115
> Optional: Transparencies of sheets 114 and 115

◊ You can start by playing the 'Three way race' several times as a class, with a track on the board.

◊ When pupils play the game themselves, ask them to record the results and then pool the class's results.

◊ Let pupils consider each other's ways of making the game fairer (if they can think of any!). Do they agree that they would be fairer?

Rat races

The 'First rat race' is straightforward (although there may be some pupils who think that 6 is 'harder' to get than other numbers). In the second race you could ask for suggestions for making it fairer, still using two dice. (For example, the track could be shortened for the 'end' numbers – even so, Rat 1 is never going to win!)

'The second race works well as a class if a rat number is assigned to a small group of pupils.'

'We had volunteer rats, a bookie, and the rest were punters as the rats moved across the room.'

For the 'Second rat race', pupils could list possible outcomes to discover that there are more ways to make 7 than there are to make 3, for example. So some scores are more likely than others.

This is taken up in section G.

C Probability (p 7)

◊ Explain first the meanings of the two endpoints of the scale. Something with probability 0 is often described as 'impossible'. However, there are different ways of being impossible and some of them have nothing to do with probability (for example, it is impossible for a triangle to have four sides). So it is better to say 'never happens'. Something with probability 1 always happens, or is certain to happen.

◊ Go through the events listed in the pupil's book and discuss where they go on the scale. The coin example leads to the other especially important point on the scale, $\frac{1}{2}$. Associate this with 'equally likely to happen or not happen', with fairness, 'even chances', etc.

◊ Keep the approach informal. The important thing is to locate a point on the right side of $\frac{1}{2}$, or close to one of the ends when appropriate (for example, in the case of the National Lottery!).

Ⓓ **Equally likely outcomes** (p 8)

◊ A spinner is very useful in connection with probability. It shows fractions in a familiar way.

D4 If the pupil's answer for (d) is $\frac{1}{3}$, then they have ignored the inequality of the parts.

Odds

Although some teachers would like to outlaw 'odds', this language is used a lot in the real world. So it may be better to explain the connection, and the difference, between probability and odds.

Bookmakers' odds make an allowance for profit and are not linked to probability in the simple way shown on the pupil's page. It is only 'fair odds' that are so linked.

Ⓔ **Equivalent fractions** (p 10)

◊ 'Pie' diagrams can be used to explain why the numerator and denominator are both multiplied by the same number. For example, in the case of $\frac{3}{4}$, each of the quarters can be subdivided into, say, 5 equal parts, giving $\frac{15}{20}$ as an equivalent fraction.

◊ You may need to emphasise that equivalence works both ways: $\frac{3}{6}$ is equivalent to $\frac{1}{2}$ and vice versa.

◊ Some pupils may have a tendency to produce a list of equivalent fractions by doubling the numerator and denominator each time, for example

$$\frac{1}{3} = \frac{2}{6} = \frac{4}{12} = \frac{8}{24} = \ldots$$

Emphasise that this strategy leads to missed fractions, for example $\frac{3}{9}$.

Ⓕ **Choosing at random** (p 12)

◊ In some cultures, raffles and all forms of gambling are disapproved of. But if there are no objections you could simulate a raffle in class.

◊ There are some misconceptions which are worth bringing into the open. Some people think that a 'special' number, like 1 or 100, is less likely to win than an 'ordinary' number (because there are fewer 'special' than 'ordinary' numbers).

F4 Part (e) assumes knowledge of factors. You may wish to check pupils' knowledge before they try this question.

F6,7 These questions should lead to discussion. Pupils may not have a strategy for comparing fractions, but may still give valid reasons for their choices. For example, 'B has twice as many reds as A but more than twice as many greens, so it's worse'.

In F7, pupils may say 'Choose D, because it has 3 more greens and only 2 more reds'. The choice is correct but the reasoning is not. Suppose, for example, bag C had 2 green and 1 red and bag D had 5 green and 3 red. The probability of choosing green from C would be $\frac{2}{3}$, and from D $\frac{5}{8}$. $\frac{2}{3}$ is greater than $\frac{5}{8}$, so C would be the better choice.

Ⓖ **Revisiting games of chance** (p 14)

Pupils list the outcomes for the throws of two dice, etc. This leads to explanations for some of the findings in section B.

Listing pairs of outcomes is covered in more depth in a later unit ('No chance').

◊ Make sure that both 1, 5 and 5, 1, for example, appear in the list of pairs.

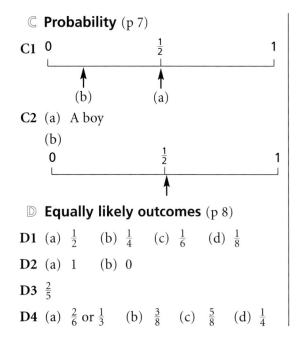

Ⓒ **Probability** (p 7)

C1

C2 (a) A boy
(b)

Ⓓ **Equally likely outcomes** (p 8)

D1 (a) $\frac{1}{2}$ (b) $\frac{1}{4}$ (c) $\frac{1}{6}$ (d) $\frac{1}{8}$

D2 (a) 1 (b) 0

D3 $\frac{2}{5}$

D4 (a) $\frac{2}{6}$ or $\frac{1}{3}$ (b) $\frac{3}{8}$ (c) $\frac{5}{8}$ (d) $\frac{1}{4}$

D5 (a) $\frac{1}{6}$ (b) $\frac{2}{6}$ or $\frac{1}{3}$ (c) $\frac{3}{6}$ or $\frac{1}{2}$

D6 (a) $\frac{5}{6}$ (b) $\frac{5}{8}$ (c) $\frac{3}{8}$ (d) $\frac{3}{4}$

D7 $\frac{3}{5}$

D8 (a) $\frac{2}{3}$ (b) $\frac{1}{8}$ (c) $\frac{4}{9}$ (d) $\frac{7}{10}$ (e) $\frac{1}{2}$

Ⓔ **Equivalent fractions** (p 10)

E1 $\frac{3}{12}$

E2 $\frac{2}{10}$

E3 $\frac{6}{8}$, $\frac{9}{12}$, $\frac{12}{16}$

E4 (a) $\frac{1}{2}$ (b) $\frac{3}{4}$ (c) $\frac{1}{4}$ (d) $\frac{3}{4}$ (e) $\frac{3}{4}$

E5 (a) $\frac{1}{4}$ (b) $\frac{3}{8}$ (c) $\frac{2}{3}$ (d) $\frac{3}{7}$ (e) $\frac{1}{5}$

E6 (a) $\frac{2}{3}$ (b) $\frac{5}{8}$ (c) $\frac{2}{5}$
(d) Cannot be simplified (e) $\frac{2}{5}$

E7 (a) $\frac{1}{3}$ (b) Cannot be simplified   (c) $\frac{2}{3}$

     (d) Cannot be simplified   (e) $\frac{2}{5}$ (f) $\frac{2}{3}$

     (g) $\frac{3}{7}$ (h) Cannot be simplified   (i) $\frac{5}{8}$

     (j) $\frac{4}{15}$

E8 $\frac{2}{3}$

F Choosing at random (p 12)

F1 $\frac{1}{50}$

F2 $\frac{4}{200} = \frac{1}{50}$

F3 $\frac{1}{25}$

F4 (a) $\frac{1}{8}$ (b) $\frac{1}{4}$ (c) $\frac{3}{8}$ (d) $\frac{5}{8}$

     (e) $\frac{1}{2}$ (f) 0 (g) $\frac{3}{4}$

F5 (a) $\frac{1}{100}$ (b) $\frac{1}{65}$ (c) $\frac{1}{64}$

F6 Sarah should choose bag A.

     $\frac{3}{8} = \frac{9}{24}$; $\frac{6}{18} = \frac{1}{3} = \frac{8}{24}$

F7 Dilesh should choose bag D.

     $\frac{4}{7} = \frac{48}{84}$; $\frac{7}{12} = \frac{49}{84}$

G Revisiting games of chance (p 14)

G1 (a) 1, 1   1, 2   1, 3   1, 4   1, 5   1, 6
       2, 1   2, 2   etc. up to 6, 6

     (b) Two even: 9 pairs
       Two odd: 9 pairs
       One even, one odd: 18 pairs

     (c) It explains why C wins most often.

G2 (a)

Total	12	11	10
Number of pairs	1	2	3

9	8	7	6	5	4	3	2
4	5	6	5	4	3	2	1

     (b) It explains why the middle numbers win more often.

G3 (a) $\frac{3}{36}$ or $\frac{1}{12}$

     (b) 7, with probability $\frac{6}{36}$ or $\frac{1}{6}$

G4 (a) 9   (b) $\frac{9}{36}$ or $\frac{1}{4}$

     (c) $\frac{9}{36}$ or $\frac{1}{4}$ (d) $\frac{18}{36}$ or $\frac{1}{2}$

G5 (a) $\frac{4}{36}$ or $\frac{1}{9}$

     (b)

Product	1	2	3	4
Probability ($\overline{36}$)	1	2	2	3

5	6	8	9	10	12	15	16	18
2	4	2	1	2	4	2	1	2

20	24	25	30	36
2	2	1	2	1

     (c) 0

G6 (a) The game is not fair.
     The probability of each difference is:

0	1	2	3	4	5
$\frac{6}{36}$	$\frac{10}{36}$	$\frac{8}{36}$	$\frac{6}{36}$	$\frac{4}{36}$	$\frac{2}{36}$

     The probability that Jack wins is $\frac{24}{36}$ and that Jill wins is $\frac{12}{36}$.

     (b) One way to modify the game is for Jack to win if the difference is odd.

G7 (a) Sc, Sc   Sc, Pa   Sc, St
       Pa, Sc   Pa, Pa   Pa, St
       St, Sc   St, Pa   St, St

     (b) $\frac{3}{9}$ or $\frac{1}{3}$

G8 All the different orders are equally likely. There are 24 of them.

     The probability that Jo will be right is $\frac{1}{24}$.

What progress have you made? (p 16)

1 (a) (see diagram: 0 ... $\frac{1}{2}$ ... 1)

     (b) It never happens.

     (c) It always happens (or it is certain).

     (d) See diagram.

2 (a) $\frac{2}{5}$ (b) $\frac{1}{4}$

3 $\frac{4}{80}$ or $\frac{1}{20}$

4 (a) $\frac{3}{5}$ (b) $\frac{5}{9}$ (c) $\frac{2}{5}$

5 (a) 1, 1   1, 2   1, 3   1, 4
       2, 1   2, 2   2, 3   2, 4
       3, 1   3, 2   3, 3   3, 4

     (b) $\frac{3}{12}$ or $\frac{1}{4}$

Practice booklet

Sections D, E and F (p 3)

1 Spinner A

   (a) $\frac{4}{8}$ or $\frac{1}{2}$ (b) $\frac{3}{8}$ (c) $\frac{1}{8}$ (d) $\frac{5}{8}$

   Spinner B

   (a) $\frac{2}{6}$ or $\frac{1}{3}$ (b) $\frac{1}{6}$ (c) $\frac{3}{6}$ or $\frac{1}{2}$ (d) $\frac{5}{6}$

   Spinner C

   (a) $\frac{1}{3}$ (b) $\frac{1}{3}$ (c) $\frac{1}{3}$ (d) $\frac{2}{3}$

2 (a) A   (b) A   (c) B

3 Three fractions equivalent to

   (a) $\frac{1}{4}$, e.g. $\frac{2}{8}$ $\frac{3}{12}$ $\frac{4}{16}$ $\frac{5}{20}$ $\frac{6}{24}$

   (b) $\frac{3}{5}$, e.g. $\frac{6}{10}$ $\frac{9}{15}$ $\frac{12}{20}$ $\frac{15}{25}$ $\frac{18}{30}$

   (c) $\frac{5}{8}$, e.g. $\frac{10}{16}$ $\frac{15}{24}$ $\frac{20}{32}$ $\frac{25}{40}$ $\frac{30}{48}$

   (d) $\frac{4}{7}$, e.g. $\frac{8}{14}$ $\frac{12}{21}$ $\frac{16}{28}$ $\frac{20}{35}$ $\frac{24}{42}$

4 (a) $\frac{1}{2}$ (b) $\frac{2}{3}$ (c) $\frac{3}{5}$ (d) $\frac{2}{3}$ (e) $\frac{7}{12}$

5 (a) $\frac{25}{55}$ Others are equivalent to $\frac{1}{2}$.

   (b) $\frac{9}{15}$ Others are equivalent to $\frac{2}{3}$.

   (c) $\frac{35}{50}$ Others are equivalent to $\frac{4}{5}$.

6 (a) $\frac{24}{36} = \frac{2}{3}$ (b) Cannot be simplified

   (c) $\frac{30}{140} = \frac{3}{14}$ (d) $\frac{8}{54} = \frac{4}{27}$

7 (a) $\frac{2}{9}$ (b) $\frac{1}{9}$ (c) $\frac{6}{9}$ or $\frac{2}{3}$

   (d) $\frac{6}{9}$ or $\frac{2}{3}$ (e) $\frac{3}{9}$ or $\frac{1}{3}$ (f) $\frac{4}{9}$

8 The probability of an orange sweet from A is $\frac{3}{9} = \frac{1}{3} = \frac{10}{30}$.

   The probability of an orange sweet from B is $\frac{6}{20} = \frac{3}{10} = \frac{9}{30}$.

   So Ann should pick a sweet from A since the probability of picking an orange sweet from that bag is greater.

9 The table shows the probabilities, and the bag from which Rick should pick to have most chance of getting a sweet he likes. Decimal probabilities have been used to help comparison, but equivalent fractions could also be used.

	Blue	Blue or yellow
Bag P	0.3	0.6
Bag Q	0.333	0.5
Bag R	0.25	0.625
Rick's choice	Bag Q	Bag R

Section G (p 5)

1 (a)

   (b) 6

   (c) $\frac{11}{36}$

2 $\frac{4}{9}$

3 $\frac{4}{9}$

4 $\frac{4}{9}$

*5 If you choose last then you have an advantage because:

   if the first player chooses A, you can pick B;
   if the first player chooses B, you can pick C;
   if the first player chooses C, you can pick A.

   In each case the probability you win is $\frac{5}{9}$.

*6 This is a table of all the possible outcomes. The winning number is ringed.

   (a) $\frac{8}{27}$ (b) $\frac{11}{27}$ (c) $\frac{8}{27}$

 **Symmetry**

Essential	Optional
Tracing paper Square dotty paper Triangular dotty paper Sheets 117 and 120 to 124	OHP transparencies made from sheets 116 and 123
Practice booklet pages 6 to 10	

Ⓐ **What is symmetrical about these shapes?** (p 17)

Discussion should show how much pupils know already about reflection and rotation symmetry.

> Optional: OHP transparency made from sheet 116, tracing paper

◊ One way to generate discussion is for pupils to study the page individually, then discuss it in small groups; then you can bring the whole class together and ask for contributions from the groups.

◊ Most pupils should be able to describe the reflection symmetry of the shapes. Some may realise that shapes with only rotation symmetry (B, C, D, I) are symmetrical in some way but be unable to describe how. Others may know about rotation symmetry already.

'Good introduction. Many children realise that shapes have symmetry but don't know why. Discussion about each of these shapes helped considerably.'

Ⓑ **Rotation symmetry** (p 18)

> Tracing paper, sheet 117
> Optional: Square dotty paper

◊ The shape on the page is the first one on sheet 117. Pupils can make and trace their own copy of this shape on square dotty paper or trace the one on the sheet.

◊ Emphasise that the shape can be rotated by putting a pencil point at the centre of rotation and turning the tracing round this fixed point.

◊ If a shape has rotation symmetry of order 1, then every point in its plane is a 'centre of rotation symmetry', because if the shape is rotated through 360° about any point it returns to its original position. Rotation symmetry of order 1 is of no interest and is ignored after question B1.

C Making designs (p 19)

> Tracing paper, square dotty paper, triangular dotty paper,
> sheets 120 and 121

D Rotation and reflection symmetry (p 21)

> Sheet 122, square dotty paper

D4 Pupils can extend this by finding all possible ways to shade four squares
to produce a design with rotation symmetry.

E Pentominoes (p 22)

> Square dotty paper

F Infinite patterns (p 23)

> Sheets 123 and 124
> Optional: two OHP transparencies of the first pattern on sheet 123,
> tracing paper

◊ OHP transparencies are ideal for showing the rotation symmetries of an
infinite pattern. Use one transparency for the fixed pattern and the other
for the tracing.

Alternatively, pupils can (if they need to) make their own tracing and use
it with the copy of the pattern on sheet 123. They only need to trace the
squares (diamonds), not the grid.

◊ The main source of difficulty is realising that both the printed pattern
and the tracing are only parts of an infinite pattern. The finite tracing
may not fit the finite part of the pattern.

These are the lines of
symmetry and centres
of rotation symmetry
(each marked with its
order of rotation
symmetry).

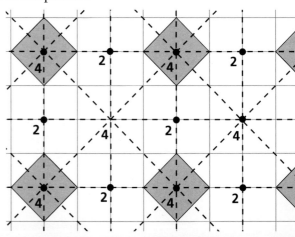

Ⓖ **Translation** (p 24)

> Sheets 123 and 124
> Optional: tracing paper

This is a brief introduction to translation. The topic will be developed further at a later stage.

◊ Make it clear that the position of the arrow representing the translation is not significant. The whole tracing is translated and the arrow merely records how far and in what direction.

G1 This question highlights a common misconception: pupils think the translation bridges the 'gap' between two shapes. If they do not see the mistake, ask them to trace the triangle and translate it 3 units across and 1 up.

Ⓑ **Rotation symmetry** (p 18)

B1 Centres of rotation marked on sheet 117

Shapes with rotation symmetry	Order of rotation symmetry
A	4
B	3
C	2
E	4
F	2
G	3
I	2
J	6
K	8
L	2
M	4

Ⓒ **Making designs** (p 19)

C1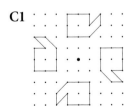

C2 Completed designs on sheet 120

C3 (a) 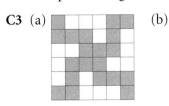 (b)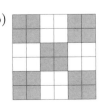

(c)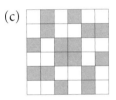

C4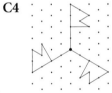

C5 Completed designs on sheet 121

C6 (a) (b) (c)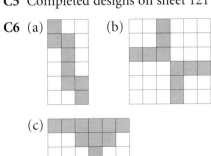

C7 (a) (b) (c)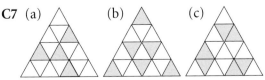

D Rotation and reflection symmetry (p 21)

D1

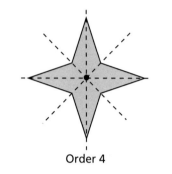

Order 4

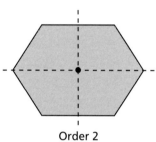

Order 2

D2 (a) Rotation symmetry of order 2
No lines of symmetry

(b) Rotation symmetry of order 3
Three lines of symmetry

(c) Rotation symmetry of order 3
No lines of symmetry

(d) No rotation symmetry
One line of symmetry

(e) Rotation symmetry of order 3
No lines of symmetry

(f) Rotation symmetry of order 2
Two lines of symmetry

(g) Rotation symmetry of order 2
No lines of symmetry

(h) Rotation symmetry of order 2
Two lines of symmetry

(i) No rotation symmetry
One line of symmetry

(j) No rotation symmetry
No lines of symmetry

(k) Rotation symmetry of order 2
No lines of symmetry

(l) No rotation symmetry
No lines of symmetry

(m) No rotation symmetry
No lines of symmetry

(n) Rotation symmetry of order 2
No lines of symmetry

D3 (a) Order 2

(b) Yes, two lines of symmetry

D4 There are sixteen ways to shade four squares to make a pattern with rotation symmetry (plus twelve that are 90° rotations of some of the sixteen). Pupils have to find eight different ways.

The sixteen ways are shown below.

Rotation symmetry of order 2

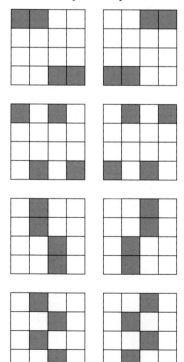

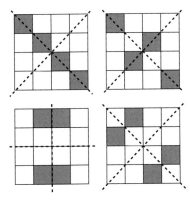

Rotation symmetry of order 4

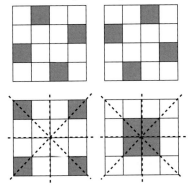

D5 (a) The pupil's pattern from

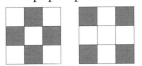

(b) The pupil's pattern from

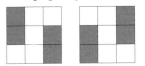

or 90° rotations of these

(c) Some examples are

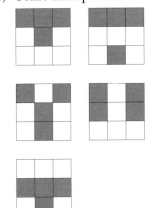

(d) Some examples are

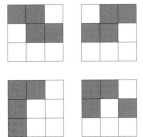

E **Pentominoes** (p 22)

E1 The pupil's pentomino from

E2 (a) The pupil's pentomino from

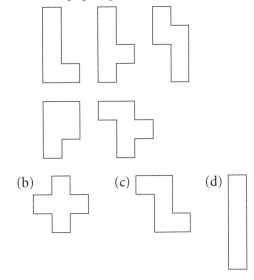

E3 (a) (i) The pupil's shape; examples are

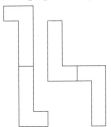

(ii) The pupil's shape; examples are

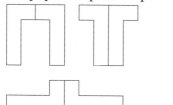

(iii) The pupil's shape; examples are

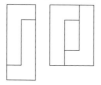

(b) The pupil's design; examples are

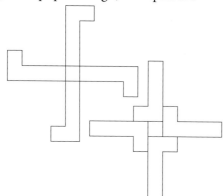

E4 The pupil's designs with

(a) reflection symmetry but no rotation symmetry

(b) rotation symmetry but no reflection symmetry

(c) reflection symmetry and rotation symmetry

F Infinite patterns (p 23)

F1

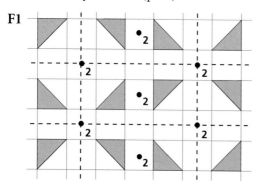

F2

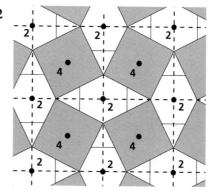

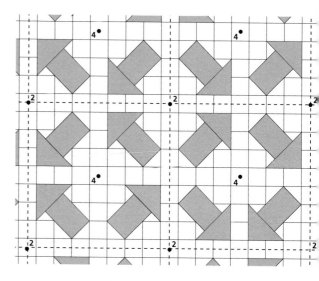

G Translation (p 24)

G1 Karl is wrong.
It should be ' 5 right, 1 up'

G2 Examples of possible translations are shown here.

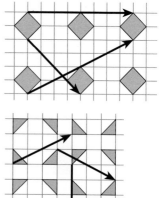

G3 Examples are shown here

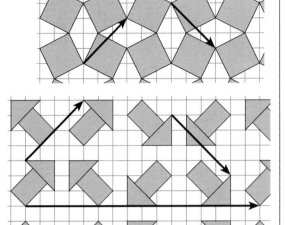

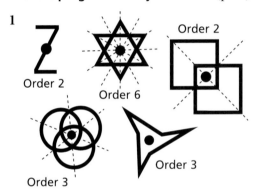

What progress have you made? (p 25)

1

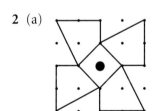

Order 2
Order 2
Order 6
Order 3
Order 3
Order 3

2 (a)

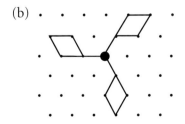

(b)

3 The pupil's pattern; examples are

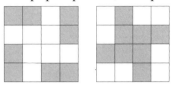

4 (a) 3 right, 1 up

(b) 3 right, 3 up; 6 left; and so on

Practice booklet

Sections A and B (p 6)

1 Designs A, C and D have rotation symmetry (of order greater than 1).

2 (a) 5   (b) 8   (c) 2   (d) 4

(e) 3   (f) 3   (g) 2   (h) 2

Section C (p 7)

1 The pupil's completed designs with rotation symmetry of order 4

2 The pupil's completed designs with rotation symmetry of order 2

3 The pupil's completed designs with rotation symmetry of order 3

Sections D and E (p 8)

1 (a) No lines of symmetry
Rotation symmetry of order 3

(b) Two lines of symmetry
Rotation symmetry of order 2

(c) One line of symmetry
No rotation symmetry

(d) No lines of symmetry
No rotation symmetry

(e) Two lines of symmetry
Rotation symmetry of order 2

(f) No lines of symmetry
Rotation symmetry of order 3

2 (a) No lines of symmetry
  Rotation symmetry of order 2

  (b) One line of symmetry
  No rotation symmetry

  (c) No lines of symmetry
  No rotation symmetry

  (d) No lines of symmetry
  Rotation symmetry of order 2

3 (a)

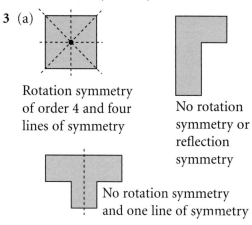

Rotation symmetry
of order 4 and four
lines of symmetry

No rotation
symmetry or
reflection
symmetry

No rotation symmetry
and one line of symmetry

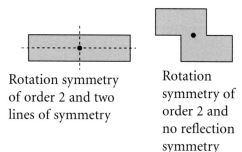

Rotation symmetry
of order 2 and two
lines of symmetry

Rotation
symmetry of
order 2 and
no reflection
symmetry

(b) It has one line of reflection
  symmetry and no rotation
  symmetry.

(c) Examples are

(i)

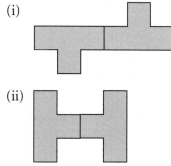

(ii)

(d) The pupil's designs with

(i) reflection symmetry but no
  rotation symmetry

(ii) reflection symmetry and
  rotation symmetry

(iii) rotation symmetry of order 4

Sections F and G (p 10)

1 3

2 4

3 No

4 A (order 2), B (order 2), D (order 2)

5 4 units right, 2 up

6 r, s, t

❺ Multiples and factors

> **Essential**
>
> Sheets 134 to 138
> Scissors
>
> **Practice booklet** pages 11 and 12

Ⓐ The sieve of Eratosthenes (p 26)

> Sheet 134

◊ Eratosthenes of Cyrene (276–194 BCE) was a Greek who is best known for measuring the Earth's circumference by observing the direction of the Sun at two places a great distance apart.

The 'sieve' is his best-known contribution to mathematics. The ringed numbers are, of course, the prime numbers.

Ⓑ Factor pairs (p 26)

Ⓒ Factor trees (p 27)

◊ If we allow, 2 to lead to the pair '2, 1' then the tree will go on for ever.

◊ It is common for pupils to forget about multiplication and write down pairs which add together to make the number.

Ⓓ Prime factorisation (p 28)

Ⓔ Lowest common multiple (p 29)

Ⓕ Highest common factor (p 30)

Ⓖ Testing for prime numbers (p 31)

Pupils should have all the prime numbers less than 100 from the sieve of Eratosthenes.

Ⓗ Clue-sharing (p 32)

> Sheets 135 to 138, scissors

◊ These puzzles are designed to encourage pupils to work collaboratively in pairs or small groups. The digits are to be cut out and put in the squares according to the clues. The clue cards are dealt out to the group members. Only the person who gets a card is allowed to see it, so they have to tell the others the clues on their cards. This makes it impossible for anyone to 'opt out', as all the clues are needed to solve the puzzle.

◊ Challenge pupils to find which puzzle has redundant clues ('Star'). Which clues are redundant? Why?

◊ After solving some of the puzzles, pupils could have a go at making up their own puzzles with clue cards.

◊ The solutions to the four puzzles are as follows.

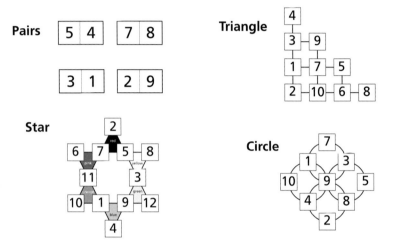

▯ **Problems** (p 32)

Ⓑ **Factor pairs** (p 26)

B1 (a) 1 2 3 4 6 12

(b) 1 2 3 6 9 18

(c) 1 2 3 4 6 8 12 24

(d) 1 2 3 5 6 10 15 30

B2 (a) 1 2 3 4 6 9 12 18 36

There is an odd number of factors. One factor pairs off with itself.

(b) Square numbers (e.g. 4, 9, 16, ...)

B3 (a) 1 _____ 17

There are only two factors, 1 and the number itself, because 17 is prime.

B4 When 10 is a factor: number ends in 0.

When 5 is a factor: number ends in 5 or 0.

When 4 is a factor: last two digits make a number which is a multiple of 4.

When 6 is a factor: number is even and 3 is a factor (using the test for 3).

When 9 is a factor, the sum of the digits is a multiple of 9.

C Factor trees (p 27)

C1 (a), (b)

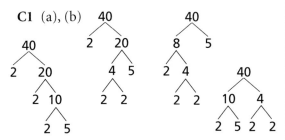

The trees all end with the same set of numbers, 2, 2, 2, 5.

C2 (a) Tree ending 2, 2, 7

(b) Tree ending 2, 3, 5

(c) Tree ending 2, 2, 2, 3, 3

(d) Tree ending 2, 2, 5, 5

C3 (a) A = 90, B = 6, C = 15

(b) P = 1620, Q = 54, R = 30, S = 6, T = 9, U = 15

(c) V = 18, W = 10, X = 9, Y = 2, Z = 3
or
V = 12, W = 15, X = 4, Y = 3, Z = 2

(d) J = 36, K = 3, L = 6, M = 2

D Prime factorisation (p 28)

D1 (a) $2 \times 2 \times 2 \times 5$ (b) $2 \times 2 \times 7$

(c) $2 \times 3 \times 5$ (d) $2 \times 2 \times 2 \times 3 \times 3$

(e) $2 \times 2 \times 5 \times 5$

D2 (a) 5^3 (b) 3^5

(c) 7^3 (d) $2^3 \times 3^2$

(e) $3^2 \times 5^3 \times 7^2$

D3 (a) 2^4 (b) 2×5^2

(c) $2^4 \times 3$ (d) $2^4 \times 5$

(e) $2^2 \times 3 \times 7$ (f) 2×3^3

D4 (a) $2^6 \times 3$ (b) 5×11^2

(c) $3^3 \times 5^2$ (d) $2^3 \times 3^2 \times 13$

E Lowest common multiple (p 29)

E1 (a) 4, 8, 12, 16, 20, 24, ...;
6, 12, 18, 24, 30, 36, ...

(b) 12

E2 (a) 40 (b) 60 (c) 60 (d) 60

E3 36

E4 $2 \times 2 \times 3 \times 7 = 84$

E5 $18 = 2 \times 3 \times 3$; $45 = 3 \times 3 \times 5$
LCM $= 2 \times 3 \times 3 \times 5 = 90$

E6 (a) 150 (b) 240 (c) 48

(d) 126 (e) 168

F Highest common factor (p 30)

F1 (a) 1, 2, 4, 8, 16; 1, 2, 4, 5, 10, 20

(b) 4

F2 (a) 4 (b) 10 (c) 15 (d) 2

F3 4

F4 $2 \times 3 \times 3 = 18$

F5 $32 = 2 \times 2 \times 2 \times 2 \times 2$
$80 = 2 \times 2 \times 2 \times 2 \times 5$
HCF $= 2 \times 2 \times 2 \times 2 = 16$

F6 (a) 15 (b) 16 (c) 8

(d) 6 (e) 1

F7 (a) 240 (b) 15 (c) 6 (d) 144

F8 (a) 180 (b) 12 (c) 14 (d) 30

F9 (a) 240 (b) 12

G Testing for prime numbers (p 31)

G1 1927 is not prime; it is divisible by 41.

G2 1931 is prime.
As you try each prime number, the other 'factor' gets smaller. When the two are as close as possible then there is no need to try any further.

G3 The statement is true for $n = $ 1, 2, 3, 4, 5, 6, ... up to 15.
But it is false for $n = 16$
$(16^2 + 16 + 17 = 17^2)$
and obviously false for $n = 17$.

G4 $(2 \times 3 \times 5 \times 7) + 1 = 211$, which is prime.
$(2 \times 3 \times 5 \times 7 \times 11) + 1 = 2311$, which is prime.

$(2 \times 3 \times 5 \times 7 \times 11 \times 13) + 1 =$
30 031, which is divisible by 59.

$(2 \times 3 \times 5 \times 7 \times 11 \times 13 \times 17) + 1 =$
510 511, which is divisible by 19.

▯ Problems (p 32)

I1 360 seconds

I2 $240 = 2 \times 2 \times 2 \times 2 \times 3 \times 5$

Any combination picked from this collection will give a factor of 240, e.g. $2 \times 2 \times 5$.

The complete set (apart from 1) is
2, 2×2, $2 \times 2 \times 2$, $2 \times 2 \times 2 \times 2$,

then each of these with $\times 3$, with $\times 5$, and with $\times 3 \times 5$, and 3, 5 and 3×5.

There are 20 factors altogether.

*I3 (a) 61 (add 1 to the LCM of 2, 3, 4, 5, and 6)

(b) 59 (The number which is 1 less than the LCM of 2, 3, 4, 5 and 6 will leave a remainder of 1 when divided by 2, of 2 when divided by 3, etc.)

The numbers are spaced at intervals of 60. There are 16 of them below 1000 (59, 119, 179, … , 959).

The problems can also be approached as follows

Numbers leaving remainder 1 when divided by 2: 1, 3, 5, 7, 9, … and so on

What progress have you made? (p 32)

1 (a) Tree ending 3, 3, 5

(b) Tree ending 2, 2, 3, 5

2 (a) $2 \times 2 \times 2 \times 2 \times 2 \times 3$

(b) $2 \times 2 \times 3 \times 5 \times 11 \times 13$

3 $2^5 \times 3$

4 56

5 15

Practice booklet

Sections B, C and D (p 11)

1 (a) $7 + 23$, $11 + 19$, $13 + 17$

(b) $3 + 97$, $11 + 89$, $17 + 83$, $29 + 71$, $41 + 59$, $47 + 53$

2 (a) Tree ending 2, 2, 3, 3

(b) Tree ending 2, 5, 7

3 (a) A is 15, B is 30, C is 6 and D is 180.

(b) A is 2, B is 60, C is 30, D is 15, E and F are 3 and 5.

4 (a) $2^2 \times 3 \times 5$

(b) $600 = 60 \times 2 \times 5 = 2^3 \times 3 \times 5^2$

5 $2^6 \times 3^2 \times 5^2$

6 (a) $a = 3$, $b = 2$ (b) $a = 2$, $b = 3$, $c = 7$

7 (a) 2^4 (b) $2^2 \times 3^2$ (c) $2^2 \times 5^2$

(d) $2^2 \times 3^2 \times 5^2$

(e) In the prime factorisation of a square number every prime factor is raised to an even power.

8 (a) $2^2 \times 3^2 \times 7 \times 23$ (b) $2 \times 3^5 \times 11$

(c) $2^4 \times 3^2 \times 53$

Sections E, F, G and I (p 12)

1 (a) 60 (b) 84

2 (a) 108 (b) 360 (c) 22 050 (d) 8100

3 24

4 (a) 6 (b) 10

(c) $3^2 \times 7 = 63$ (d) $2^2 \times 5^2 = 100$

5 23

6 They will ring again when the time passed in minutes is equal to the LCM, that is 288 minutes later, or 4:48 p.m.

7 The jug size will be the HCF of 72, 24, 56 and 120.
This is 8 litres.

8 101, 131 and 151

Number grids

In this unit, pupils solve number grid problems using addition and subtraction. This includes using the idea of an inverse operation ('working backwards').

Algebra arises through investigating number grids. Pupils simplify expressions such as $n + 4 + n - 3$ and produce simple algebraic proofs of general statements. They also simplify expressions such as $2n \times n$.

Optional
A4 sheets of paper
Felt-tip pens or crayons

Practice booklet pages 13 to 16

A Square grids (p 33)

The idea of a number grid is introduced. There are many opportunities to discuss mental methods of addition and subtraction.

Optional: A4 sheets of paper and felt-tip pens or crayons (for 'Human number grids')

T

Human number grids

This introductory activity does not appear in the pupil's book.

◊ Each pupil or pair of pupils represents a position in a number grid.

Each position will contain a number. (For pupils familiar with spreadsheets, the idea of a 'cell' may help.)

The operations used are restricted to addition and subtraction.

◊ Tables/desks need to be arranged in rows and columns so that the cells form a grid. Explain, with appropriate diagrams, that the class is going to form a human number grid that uses rules to get from a number in one cell to a number in another. A possible diagram is shown below.

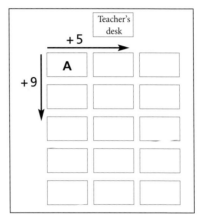

◊ Referring to the numbers is easier if the cells are labelled.
Pupils can discuss how each cell might be labelled, for example:

- A1, A2, B1, ... as on a spreadsheet

- A, B, C, ...

- or with the pupils' names

◊ Initially, it may help to use only addition or use sufficiently large numbers in cell A to avoid the complication of negative numbers.

◊ Decide on the first pair of rules and ask the pupils in the cell marked A in the diagram to choose a number for that cell.
Discuss how the numbers in other cells are found.
Now ask the pupils in cell A to choose another number, write it on both sides of a sheet of paper and hold it up.
Pupils now work out what number would be in their cell, write it on both sides of their sheet of paper and hold it up.
This can be repeated with different pupils deciding on the number for their cell.

◊ Questions can be posed in a class discussion, for example:

  • Suppose the number in Julie and Asif's cell is 20.
    What number is in your cell, Peter?
    What number will be in Jenny's cell?

  • What number do we need to put in cell A so that
    the number in cell F is 100?

  • Find a number for cell A so that the number in cell K is negative.

  • What happens if the 'across' and 'down' rules change places?

Ask pupils to explain how they worked out their answers. You could
introduce the idea of an 'inverse' and encourage more confident pupils to
use this word in their explanations.

Square grids

◊ Point out that all grids in the unit are square grids.

◊ One teacher presented unfinished grids on an OHP transparency and
asked for volunteers to fill in any empty square. She found that less
confident pupils chose easy squares to fill in while 'others with more
confidence chose the hardest, leading to class discussion, and the idea of a
"diagonal" rule came out naturally.'

◊ In one school, the class looked at rules in every
possible direction as shown in the diagram.

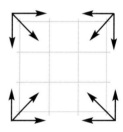

◊ In discussion, bring out the fact that there are different ways to calculate a
number in a square depending on your route through the grid.

A1 In part (b), make sure pupils realise that the diagonal rule fits **any**
position on these grids and not just those on the leading diagonal.
In part (c), some pupils may continue to use the '+ 6' and '+ 2' rules here.
Discuss why using the diagonal rule '+ 8' could be used to give the same
result.

A3 Some could consider what the diagonal rule is if the across and down
rules are '+ a' and '+ b' or '– x' and '– y'.

A4 Some pupils can look for all possible pairs, introducing the idea of an
infinite number of pairs of the forms

'+ a' and '+ $(11 - a)$' in part (a)

'+ b' and '+ $(4 - b)$' in part (b)

B Grid puzzles (p 34)

◊ Pupils need not complete the grids to solve the puzzles; just find the missing number or rules.

◊ Encourage pupils to use the word 'inverse' when describing their methods.

B3 Pupils could solve puzzle (c) using trial and improvement or possibly by the following more direct method.

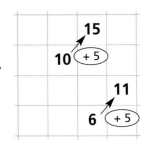

The rule in this diagonal direction (↗) is '+5' so the number in the top row directly above the 6 is 10 + 5 = 15.

Now the down rule is easily found to be '–3' and the across rule is '+2'.

As a possible extension, pupils could make up their own puzzles.

*B5 Pupils can devise their own methods to solve these. Ask them to explain them to you or each other. Some may use a direct method such as the one above. Others may devise systematic trial and improvement methods.

C Algebra on grids (p 35)

◊ The first grid on the page uses addition only. After discussing it, you may wish pupils to try questions C1 to C3 where the rules are restricted to addition. Then discuss the second grid at the top of the page before moving on to question C4.

You may find a number line is helpful in getting across the fact that ⁻4 + 7 is equivalent to + 3, for example.

Emphasise that the expressions in the grid show how to find any number on the grid **directly** from the top left corner.

In the second grid, pupils who suggest '$h – 9$' for the square below '$h – 2$' are possibly thinking of '$h – 2 + 7$' as '$h – (2 + 7)$'. Discussion of numerical examples may help to clear up any confusion.

Ensure that pupils understand, for example, that '$n + 7 – 4$' gives the same result as '$n – 4 + 7$'.

D Grid investigations (p 37)

D3 This provides an opportunity for pupils to choose to use algebra for themselves to explain their findings.

D4 After pupils have investigated opposite corners, draw their conclusions together in a discussion that leads to the algebraic ideas in section E.

E Using algebra (p 38)

◊ This section follows on directly from the pupils' investigations in section D.
Extend your discussion to consider how to simplify expressions
such as $n - 3 + n - 4$ and $n + n + 2 + n - 3$.

Extension Pupils could find and prove that the opposite corners total on a 3 by 3 grid
is 2 times the centre number or that the diagonals total is 3 times the centre
number.

F Using multiplication (p 40)

◊ Pupils simplify expressions such as $2n \times n$ and $3x \times 2x$.

F1 Pupils have the opportunity to use algebra to explain their answer, possibly
stating that $3(n + 2) \neq 3n + 2$.
Encourage pupils to give explanations in a variety of ways using numbers,
diagrams and/or algebra.

F2 Ask pupils if it will always be possible to make a grid with two '×' rules.

F5 Pupils could substitute various values for y to convince themselves that
Tom is correct.

F12 Pupils can show algebraically that, on these grids, adding pairs of numbers
in opposite corners does not produce the same results. Make sure they find
that multiplying gives equal results and that they explain this algebraically.

A Square grids (p 33)

A1 (a) The pupil's grids
(b) The rule is '+ 8', with the pupil's
explanations.

(c)
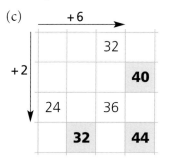

A2 (a) (i)
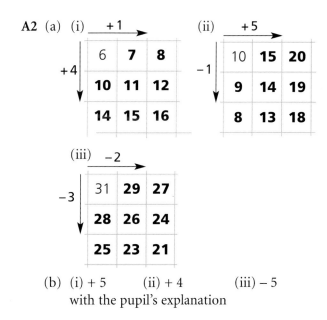

(b) (i) + 5   (ii) + 4   (iii) − 5
with the pupil's explanation

A3 (a) $+17$ (b) $+5$

A4 The pupil's pairs of rules, for example,

(a) Across '$+1$', down '$+10$', or across '$+3$', down '$+8$'

(b) Across '$+1$', down '$+3$', or across '-1', down '$+5$'

Rules in (a) are of the form $+a, +(11-a)$

Rules in (b) are of the form $+a, +(4-a)$

⒝ Grid puzzles (p 34)

B1 (a) 30 (b) 43 (c) 95

B2 (a) -3 (b) $+21$ (c) -5

B3 (a) Across '$+7$', down '-3'

(b) Across '-2', down '$+11$'

(c) Across '$+2$' down '-3'

B4 (c) is usually the most difficult. The pupil's reasons

*B5 (a) Across '-1', down '$+5$'

(b) Across '-3', down '-4'

(c) Across '$+4$', down '-1'

(d) Across '$+6$', down '-2'

⒞ Algebra on grids (p 35)

C1 (a)

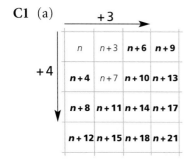

(b)

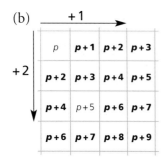

(c)

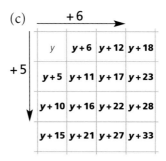

C2 (a) 121 (b) 109 (c) 133

C3 (a) Across '$+6$', down '$+5$'

(b) Across '$+5$', down '$+1$'

(c) Across '$+3$', down '$+10$'

C4 (a)

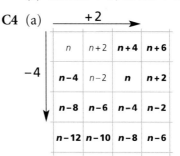

(b)

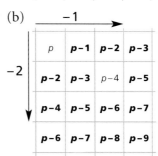

(c)

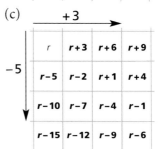

C5 (a) 56 (b) 59 (c) 56

C6 (a) $f+11$ (b) $y+12$ (c) $x+6$

(d) $z-10$ (e) $p+1$ (f) $m-3$

(g) $q+3$ (h) $w-14$ (i) $h+5$

C7

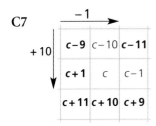

	−1 →		
+10 ↓	c−9	c−10	c−11
	c+1	c	c−1
	c+11	c+10	c+9

C8 (a)

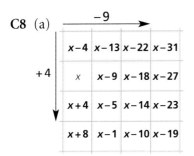

	−9 →			
+4 ↓	x−4	x−13	x−22	x−31
	x	x−9	x−18	x−27
	x+4	x−5	x−14	x−23
	x+8	x−1	x−10	x−19

(b)

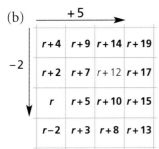

	+5 →			
−2 ↓	r+4	r+9	r+14	r+19
	r+2	r+7	r+12	r+17
	r	r+5	r+10	r+15
	r−2	r+3	r+8	r+13

(c)

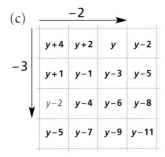

	−2 →			
−3 ↓	y+4	y+2	y	y−2
	y+1	y−1	y−3	y−5
	y−2	y−4	y−6	y−8
	y−5	y−7	y−9	y−11

C9 (a) Across '+ 4', down '− 1'

(b) Across '− 5', down '− 3'

(c) Across '− 2', down '− 1'

*C10 (a) $+ 5 − x$ or equivalent

(b) $+ 10 − n$ or equivalent

D Grid investigations (p 37)

D1 (a) $37 + 13 = 50$

(b) The pupil's grids

(c) For each grid, the opposite corners totals are the same – this result could be explained using algebra.

D2 The pupil's investigation

D3 (a)

Opposite corners table	
Top left number	Opposite corners total
2	10
3	12
4	14
10	26

(b) The pupil's grids and results

(c) 'The opposite corners total is (the top left number × 2) + 6' or '… (the top left number + 3) × 2' or '… $2n + 6$' or equivalent.

(d) 206

D4 The pupil's investigation

E Using algebra (p 38)

E1 Grid P

(a)

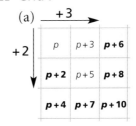

	+3 →		
+2 ↓	p	p+3	p+6
	p+2	p+5	p+8
	p+4	p+7	p+10

(b) Both pairs of corners add up to $2p + 10$.

(c) Yes

(d) 210

Grid N

(a)

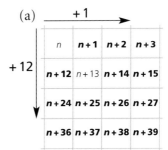

	+1 →			
+12 ↓	n	n+1	n+2	n+3
	n+12	n+13	n+14	n+15
	n+24	n+25	n+26	n+27
	n+36	n+37	n+38	n+39

(b) Both pairs of corners add up to $2n + 39$.

(c) Yes

(d) 239

Grid T

(a)

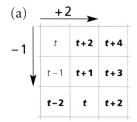

(b) Both pairs of corners add up to
$2t + 2$.

(c) Yes

(d) 202

E2 A and I, B and G, C and E, D and H

E3 The pupil's explanation, for example, the
expressions have the same value for **one**
value of n but this does not mean that
the expressions have the same value for
all values for n.

E4 (a) $2p + 6$ (b) $2y + 9$ (c) $3q + 8$
    (d) $3t + 4$ (e) $2x + 1$ (f) $3r + 6$
    (g) $2w - 9$ (h) $2j - 1$ (i) $3h - 2$

E5 The pupil's investigation

E6 Pairs that add to give $+ 15$, for example
'$+ 8$' and '$+ 7$', '$+ 16$' and '$- 1$'

F Using multiplication (p 40)

F1 (a) There is no single right answer.
They have taken different routes
through the grid.

    (b) No, it is not possible – the pupil's
explanation

F2 (a)

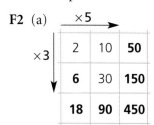

(b) The pupil's explanation

F3 (a) No   (b) Yes   (c) No
    (d) Yes   (e) Yes

F4 (a)

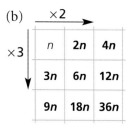

(b)

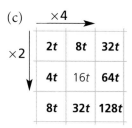

(c)

$\times 4$		
$2t$	$8t$	$32t$
$4t$	$16t$	$64t$
$8t$	$32t$	$128t$

with $\times 2$ down the side

F5 Tom is correct. The pupil's reasons

F6 (a) $2a$ (b) $2a^2$ (c) $6a^2$ (d) $9a^2$
    (e) $5x^2$ (f) $10x^2$ (g) $25x^2$ (h) $36x^2$

F7 A and D, B and J, C and F,
E and H, G and I

F8 (a) $6p^2$ (b) $5p$ (c) $6p$
    (d) $3t^2$ (e) $15t^2$ (f) $15t$
    (g) $5t^2$ (h) $24x^2$ (i) $24x^2$
    (j) $20y^2$ (k) $15y^2$ (l) $24y^2$

F9 The pupil's explanation

F10 (a) $12p$ (b) $7p$ (c) $12p^2$
    (d) Cannot be simplified
    (e) $p + 7$ (f) $2p + 7$ (g) $2p + 1$
    (h) Cannot be simplified
    (i) $8m^2$ (j) $9m$
    (k) Cannot be simplified   (l) $7m + 2$

F11 (a) $2x - 6$ (b) $16 + 3y$ (c) $6w^2$
    (d) $4u + 4$ (e) Cannot be simplified
    (f) $18p^2$ (g) Cannot be simplified
    (h) $2r - 4$ (i) $4s - 3$

F12 The pupil's investigation

What progress have you made? (p 42)

1. (a) Across '– 5', down '– 3'
   (b) Across '+ 2', down '– 3'

2. (a) $n + 7$ (b) $p + 6$ (c) $y - 7$
   (d) $t - 3$ (e) $2h + 3$ (f) $3w - 11$

3. (a)

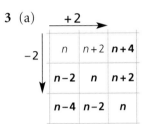

   (b) The pupil's explanation, for example, the expressions in one diagonal are all the same (n) so the numbers will be the same.
   (c) $2n$
   (d) 12
   (e) The pupil's investigation

4. (a) t^2 (b) $5g^2$
   (c) $12n^2$ (d) $10y^2$

Practice booklet

Sections A and B (p 13)

1. (a) (i)

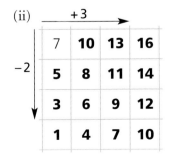

   (ii)

(iii)

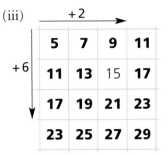

(iv)

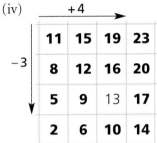

(b) (i) + 3   (ii) + 1   (iii) + 8   (iv) + 1

2. (a) + 13   (b) + 2

Section C (p 13)

1. (a) (i)

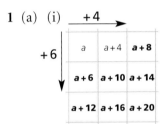

   (ii)

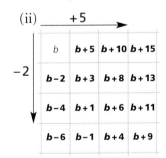

(b) (i) 120   (ii) 109
(c) (i) 80   (ii) 91

2. (a) Across '+ 3', down '+ 4'
   (b) Across '+ 5', down '+ 3'
   (c) Across '– 1', down '+ 4'

3 (a) $t + 11$ (b) $a + 7$ (c) $q + 6$

   (d) $p + 4$ (e) $x + 10$ (f) $y + 2$

   (g) $s - 4$ (h) $v - 5$ (i) $b + 6$

   (j) $a - 4$ (k) $f - 16$ (l) $c + 2$

   (m) $d - 11$ (n) $g - 2$ (o) $h - 14$

Sections D, E and F (p 14)

1 (a)

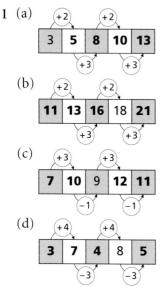

(b)

(c)

(d)

2 (a) $+ 4$ (b) $+ 7$ (c) $+ 8$ (d) $+ 4$

3 The pupil's pairs of rules, for example,
Top '+ 1', bottom '+ 3'
Top '+ 3', bottom '+ 1'
Top '+ 2', bottom '+ 2'
Top '+ 5', bottom '– 1'

4 (a)

(b)

	End total table	
	First number	End total
	3	20
	4	22
	5	24
	n	$2n + 14$

(c) 'The end total is
2 times the first number add 14' or
'… (the first number $\times$ 2) + 14' or
'… (the first number + 7) $\times$ 2' or
'… $2n + 14$ where n is the first number' or equivalent.

(d) 30

5 The pupil's investigation

6 A and D, C and I, E and G

7 (a) $2p + 3$ (b) $3y + 9$ (c) $2q + 10$

   (d) $2t + 3$ (e) $2x + 2$ (f) $3r + 6$

   (g) $2w - 13$ (h) $2j - 3$ (i) $2h - 6$

8 (a)

Grid total table	
Top left	Grid total
2	63
3	72
4	81

(b) 'The grid total is 9 times the top left number add 45' or
'… (the top left number $\times$ 9) + 45' or
'… (the top left number + 5) $\times$ 9' or
'… $9n + 45$ where n is the top left number' or equivalent.

(c) 495

9 The pupil's investigation

10 (a) $6n^2$ (b) $35p$ (c) $4m - 3$

    (d) Cannot be simplified

    (e) $3a^2$ (f) $s + 5$

7 Constructions

Ⓐ **Right angles** (p 43)

Set square

Ⓑ **From a point to a line** (p 45)

Set square, sheet 139, squared paper

The fact that the shortest distance is perpendicular to the line appears obvious when the point is above a horizontal line. You could start with this: and rotate to this:

Ⓒ **Constructions with ruler and compasses** (p 46)

Compasses

Ask pupils why the constructions work. The most accessible explanations are based on reflection symmetry.

Ⓐ **Right angles** (p 43)

A1 *a* and *h*, *b* and *d*, *e* and *g*

A2 (a) North   (b) North-east
    (c) North-east

A3 The pupil's drawing

A4 *a* and *b*, *c* and *j*, *d* and *e*, *i* and *l*.

Ⓑ **From a point to a line** (p 45)

B1 The pupil's mirror images

B2 The pupil's diagram; (4, 2)

B3 The total of the three distances is the same, whatever the position of P.

ℂ **Constructions with ruler and compasses** (p 46)

C1 The three lines (possibly extended) all go through a point.

C2 They all go through a point.

C3 They meet at the midpoint of AC.

C4 They all go through a point.

C5 The angle bisectors are at right angles.

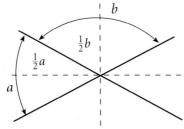

$\frac{1}{2}a + \frac{1}{2}b = 90°$ because $a + b = 180°$.

C6 The angles of the large triangle are equal to those of the original triangle.

What progress have you made? (p 49)

1 The pupil's construction

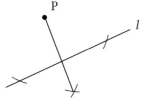

2 The pupil's construction

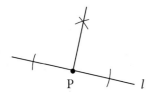

3 The pupil's construction

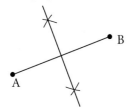

4 The pupil's construction

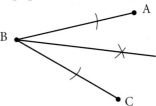

Practice booklet

Section C (p 17)

1 The pupil's construction of an angle of 45°

2 (a) 60°

   (b) The pupil's construction of an angle of 30°

3 The three points A, B, C are on a straight line.

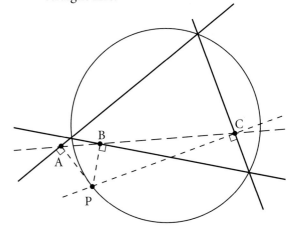

Comparisons

This unit introduces median and range and uses these to compare sets of data. Pupils generate their own data in a reaction time experiment and use it to make comparisons. Other data handling projects are suggested and there is advice on writing up.

Essential	Optional
Sheet 141	Sheets 142 to 145
Practice booklet pages 18 to 20	

⒜ **Comparing heights** (p 50)

◊ The discussion here is intended to be open, with no particular method of comparison preferred. The important thing is to give reasons for decisions.

The two questions (about the picture and about the dot plots) can be given to pairs or small groups to consider before a general discussion.

There may not always be a clear-cut answer.

◊ Pupils may suggest finding the mean. You could offer data where the mean of one group is greater than the mean of another, but only because of one extreme value (for example, 190, 150, 150, 150 and 159, 159, 159, 159). Work on the mean and discussion of which 'average' is more appropriate are dealt with later in the course.

B **Median** (p 52)

◊ You can introduce the median using the pupils' own heights. They will need to know their heights or will need to measure them.

Start by getting an odd number of pupils to stand in order of height. Emphasise that it is not the middle person who is the median of the group, but that person's height. It is a good idea to use 'median' only as an adjective at first, for example 'median height', 'median age'.

Then do the same with an even number of pupils.

To emphasise the value of the median as a way of comparing data, you may wish to carry out this activity with two separate groups (boys and girls or sides of the class).

C **Range** (p 54)

◊ The range can be shown practically with a group of pupils. Ask the group to stand in height order. Ask the tallest and shortest to stand side by side. Measure the difference between their heights.

D **How fast do you react?** (p 56)

Pupils work in pairs.

Sheet 141

◊ In addition to comparing performance within each pair, pupils could compare left and right hands. For homework they could compare themselves with an adult.

◊ In one class pupils felt that they were getting clues from twitching fingers just before the ruler was dropped, so a card was used to cover the fingers.

◊ You may wish to work through section F (on report writing) first and get pupils to write up their work on reaction times. If so, you will need to explain the diagrams in section E as well.

E **Summarising data** (p 57)

The diagram in the pupil's book is a simplified version of the 'box and whisker' diagram, which shows the quartiles as well.

Lowest   Lower quartile   Median   Upper quartile   Highest

Either type of diagram may be arranged vertically.

Make sure pupils align the diagrams for each question correctly on a common scale, so that they can be used to make comparisons.

𝔽 **Writing a report** (p 58)

The specimen report is intended to help pupils write their own reports. It can be used for either group or class discussion

◊ You may need to make it clear that the fastest person is the one with the shortest time, and vice versa.

𝔾 **Projects** (p 60)

Each project generates data for making comparisons in a short written report.

The Argon Factor

Optional: Sheets 142 and 143 (preferably on OHP transparencies), sheet 144

◊ There are two tests: mental agility and memory.

◊ In the mental agility test, give the pupils one minute to memorise the shapes and numbers. (They are best shown on an OHP.)

Tell pupils 'You will be given 5 seconds to answer each question. Questions will be read twice.'

1 What number is inside the circle?
2 What number is inside the pentagon?
3 What number is inside the first shape?
4 What shape has the number 17 in it?
5 What shape has the number 29 in it?
6 What number is inside the middle shape?
7 What number is inside the last shape?
8 What shape is in the middle?
9 What is the shape before the end one?
10 What number is in the second shape?
11 What shape is the one after the one with 21?
12 What number is inside the fourth shape?
13 What number is in the shape just right of the triangle?
14 What shape is two to the left of the kite?
15 What number is in the shape two after the circle?

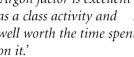

'Argon factor is excellent as a class activity and well worth the time spent on it.'

◊ In the memory test, give the pupils two minutes to remember the pictures and details of the four people on sheet 143.

Then give them ten minutes to write answers to the 20 questions on sheet 144.

◊ Discuss with the class what comparisons can be made from the scores. Suggestions include these.

- Do people remember more about their own sex?
- Which test did the class do better on? (Remember that the number of questions is not the same.)

Other projects

Optional: For 'Tile pattern' (see below), sheet 145

Another project idea is given on p 60 ('Handwriting size').

Another possibility is 'Tile pattern', for which sheet 145 is needed. This is suitable for an individual pupil or a pair. The pupil(s) carrying out the project cut out the 16 tiles and put them in an envelope. They decide which groups of people are to be compared (for example, children and adults). Each 'subject' is then timed making the pattern shown on the sheet.

⊞ Puzzles and problems (p 60)

Some questions here and in the practice leaflet involve finding the median of a frequency distribution. This topic is developed further at a later stage.

⊟ Median (p 52)

B1 (a) 159 cm   (b) 156 cm   (c) 154.5 cm

B2 (a) 11   (b) 154 cm

B3 A   Girls 136 cm; boys 153 cm
The boys are taller.

B   Girls 149.5 cm; boys 143 cm
The girls are generally taller.

C   Girls 143 cm; boys 149 cm
The boys are generally taller but the girls' heights are well spread out.

D   Girls 149 cm; boys 140 cm
The girls are generally taller but there are some short girls and one tall boy.

E   Girls 139 cm; boys 149 cm
The boys are generally taller but there is a tall girl and some short boys.

F   Girls 152 cm; boys 145 cm
The girls are generally taller.

B4 (a) 152 cm   (b) 151 cm
(c) 153 cm   (d) 152.5 cm

B5 (a) 69 kg   (b) No change
(c) No change   (d) Up by 1 kg

B6 (a) 52, 54, 58, 60, 63, 65, 70
(b) 60 kg

B7 (a) 152 cm (lengths in order are 139, 148, 152, 156, 161 cm)

   (b) 36 kg (weights in order are 26, 29, 31, 34, 38, 39, 40, 45 kg)

B8 Boys have the greater median weight (2.7 kg); the girls' median is 2.65 kg.

ℂ Range (p 54)

C1 (a) A 13 cm   B 11 cm   C 18 cm

   (b) C

   (c) B

C2 (a) 13 minutes   (b) 4 minutes

   (c) 9 minutes

C3 (a) Herd B

   (b) Herd A,
       because it has the greatest range

   (c) Herd C,
       because it has the smallest range

C4 (a) Median 28, range 12

   (b) Median 85, range 29

   (c) **Nicky** and **Carol** both had high scores, but **Nicky**'s scores were the more spread out of the two.

   (d) Nicky's scores and **Martin**'s scores were both spread out, but **Nicky** had the higher scores of the two.

   (e) **Paul** and **Martin** were both bad players because they had **low** median scores.

   (f) Paul was a consistent player because the range of his scores was **low**.

C5 (a) Northern: median 12 m, range 7 m
       Southern: median 14.5 m, range 11 m

   (b) Southern trees are taller. Their heights are more spread out.

C6 (a) Machine A: median 500 g, range 26 g
       Machine B: median 498 g, range 5 g
       Machine C: median 502 g, range 6 g
       Machine D: median 514 g, range 25 g

   (b) Machine B

   (c) Machine D

   (d) Machine C

   (e) (Machine A) Inconsistent: it both underfilled and overfilled packs

   (f) Machine C. It usually put enough in a pack to avoid complaint without being too generous to the customer.

𝔼 Summarising data (p 57)

E1 The pupil's diagrams

E2 (a)

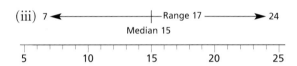

   (b) Linford is consistently quick. He has the lowest median time and the smallest range.
       Jules has some quick times, but is the least consistent, as shown by the large range.

ℍ Puzzles and problems (p 60)

H1 The missing age could be 5 or 48.
    If it is 5, the median is 16; if it is 48, the median is 20.

H2 (a) 20

   (b) (i) 11, 15 or 18

      (ii) 20

      (iii) 24, 25 or 27

H3 (a) 2 people   (b) 3 people

*H4 43, 46, 55, 68, 72 and 48, 58, 62, 79

What progress have you made? (p 61)

1 Right hand: median 14, range 6
Left hand: median 18, range 16

Pat is faster and less variable using his right hand.

Practice booklet

Sections A, B and C (p 18)

1 (a) 138 g   (b) 140 g   (c) 207 g

2 (a)

(b) 132 g

3 (a) Barcelona 21°C
Birmingham 22°C

(b) Birmingham

(c) Barcelona 12 degrees
Birmingham 8 degrees

Sections C and E (p 19)

1 The lengths are
A 27 mm   B 33 mm   C 36 mm
D 40 mm   E 56 mm   F 53 mm
Median length 38 mm
Range of lengths 29 mm

2 (a) Median 85 kg, range 46 kg

(b) Median 7.8 m, range 3.3 m

(c) Median 1°C, range 10 degrees

(d) Median 37.7°C, range 2.8 degrees

3 The pupil's comparison, such as 'The grey squirrels weigh more, because the median weight of the red squirrels is 293 g, and the median of the greys is 599 g. The greys are also more varied in weight, as the range of the reds is 25 g, but the range of the greys is 112 g.'

4 On the whole, Jo is faster (the medians are 80.1 s and 81.7 s). Jo is also more consistent (the ranges are 3.4 s and 8.3 s). But Jay has had the fastest single run (76.2 s).

5 Jo

78.3 ◄ Range 3.4 ├──────► 81.7
Median 80.1

Jay

76.2 ◄── Range 8.3 ──┤──────► 84.5
Median 81.7

Section H (p 20)

1 24

2 (a) 23   (b) 25

3 $23\frac{1}{2}$

4 2

5 17

⑨ Fractions

This unit revises earlier work on equivalent fractions and simplifying and extends to improper fractions and mixed numbers and to comparing, adding and subtracting fractions.

Practice booklet page 21

Ⓐ Equivalent fractions (p 62)

◊ The questions include examples of changing decimals to fractions.

Ⓑ Using equivalent fractions (p 64)

Ⓐ Equivalent fractions (p 62)

A1 (a) $\frac{1}{8} = \frac{2}{16}$ (b) $\frac{4}{5} = \frac{12}{15}$ (c) $\frac{2}{3} = \frac{12}{18}$

   (d) $\frac{3}{8} = \frac{12}{32}$ (e) $\frac{25}{40} = \frac{5}{8}$ (f) $\frac{16}{24} = \frac{2}{3}$

   (g) $\frac{18}{30} = \frac{6}{10}$ (h) $\frac{45}{75} = \frac{3}{5}$

A2 $\frac{9}{24} = \frac{3}{8} = \frac{6}{16}$ $\frac{8}{10} = \frac{4}{5} = \frac{16}{20}$ $\frac{2}{3} = \frac{10}{15} = \frac{8}{12}$

A3 (a) $\frac{2}{3}$ (b) $\frac{1}{2}$ (c) $\frac{3}{4}$ (d) $\frac{1}{2}$

   (e) $\frac{6}{7}$ (f) $\frac{2}{5}$ (g) $\frac{2}{7}$ (h) $\frac{4}{5}$

   (i) $\frac{2}{3}$ (j) $\frac{1}{5}$

A4 (a) $\frac{1}{3}$ (b) $\frac{4}{9}$ (c) $\frac{3}{5}$ (d) $\frac{2}{3}$

   (e) $\frac{1}{3}$ (f) $\frac{9}{16}$ cannot be simplified

   (g) $\frac{4}{5}$ (h) $\frac{3}{4}$ (i) $\frac{3}{7}$ (j) $\frac{5}{12}$

A5 (a) $\frac{1}{12}$ (b) $\frac{1}{4}$ (c) $\frac{1}{3}$ (d) $\frac{7}{16}$

   (e) $\frac{5}{8}$ (f) $\frac{35}{48}$ (g) $\frac{5}{6}$

A6 $\frac{3}{20}$

A7 (a) $\frac{6}{25}$ (b) $\frac{3}{50}$ (c) $\frac{19}{20}$ (d) $\frac{33}{100}$

   (e) $\frac{8}{25}$ (f) $\frac{4}{5}$ (g) $\frac{9}{20}$ (h) $\frac{3}{25}$

   (i) $\frac{17}{20}$ (j) $\frac{1}{50}$

A8 $\frac{33}{200}$

A9 (a) $\frac{11}{40}$ (b) $\frac{1}{8}$ (c) $\frac{37}{250}$

   (d) $\frac{11}{200}$ (e) $\frac{1}{125}$

A10 (a) fiftieths   (b) twenty-fifths

   (c) fortieths

Ⓑ Using equivalent fractions (p 64)

B1 (a) $\frac{3}{8}$ (b) $\frac{2}{3}$ (c) $\frac{4}{5}$ (d) $\frac{5}{8}$

   (e) $\frac{4}{9}$ (f) $\frac{3}{8}$ (g) $\frac{3}{5}$ (h) $\frac{9}{20}$

B2 (a) $\frac{4}{3}$ (b) $\frac{9}{4}$ (c) $\frac{17}{5}$

   (d) $\frac{17}{10}$ (e) $\frac{23}{8}$

B3 (a) $2\frac{1}{2}$ (b) $1\frac{2}{3}$ (c) $1\frac{3}{4}$

   (d) $2\frac{1}{5}$ (e) $2\frac{3}{5}$

B4 (a) $\frac{5}{6}$ (b) $\frac{7}{12}$ (c) $\frac{8}{15}$ (d) $\frac{5}{8}$

   (e) $\frac{19}{24}$ (f) $\frac{31}{40}$ (g) $1\frac{3}{8}$ (h) $1\frac{5}{12}$

   (i) $\frac{19}{24}$ (j) $1\frac{1}{20}$

B5 (a) $\frac{1}{6}$ (b) $\frac{1}{12}$ (c) $\frac{2}{15}$ (d) $\frac{3}{8}$

   (e) $\frac{7}{24}$ (f) $\frac{9}{40}$ (g) $\frac{9}{40}$ (h) $\frac{7}{20}$

   (i) $\frac{11}{24}$ (j) $\frac{1}{12}$

B6 (a) $2\frac{2}{5}$ (b) $6\frac{2}{5}$ (c) $3\frac{6}{8}$ or $3\frac{3}{4}$

   (d) $5\frac{5}{6}$ (e) $2\frac{4}{8}$ or $2\frac{1}{2}$

What progress have you made? (p 65)

1 (a) $\frac{4}{5}$ (b) $\frac{3}{4}$

2 $\frac{5}{8}$ ($\frac{5}{8} = \frac{35}{56}$, $\frac{4}{7} = \frac{32}{56}$)

3 (a) $\frac{13}{5}$ (b) $9\frac{1}{3}$

4 (a) $\frac{29}{40}$ (b) $\frac{5}{12}$ (c) $1\frac{7}{12}$

5 (a) $5\frac{1}{4}$ (b) $7\frac{1}{5}$

Practice booklet

Section A (p 21)

1 (a) $\frac{1}{4} = \frac{8}{32}$ (b) $\frac{2}{3} = \frac{14}{21}$ (c) $\frac{3}{5} = \frac{27}{45}$

2 (a) $\frac{3}{7}$ (b) $\frac{5}{8}$ (c) $\frac{5}{8}$
   (d) $\frac{1}{5}$ (e) $\frac{1}{3}$

3 (a) $\frac{11}{25}$ (b) $\frac{17}{50}$ (c) $\frac{1}{25}$
   (d) $\frac{11}{20}$ (e) $\frac{3}{10}$

Section B (p 21)

1 (a) $\frac{2}{3}$ (b) $\frac{3}{7}$ (c) $\frac{5}{6}$

2 (a) $\frac{6}{5}$ (b) $\frac{11}{8}$ (c) $\frac{7}{3}$
   (d) $\frac{23}{8}$ (e) $\frac{13}{4}$

3 (a) $1\frac{1}{2}$ (b) $2\frac{2}{5}$ (c) $3\frac{3}{4}$
   (d) $2\frac{2}{3}$ (e) $3\frac{1}{3}$

4 (a) $\frac{9}{20}$ (b) $\frac{23}{24}$ (c) $\frac{17}{21}$ (d) $\frac{31}{40}$
   (e) $1\frac{1}{12}$ (f) $1\frac{1}{10}$ (g) $1\frac{11}{24}$ (h) $1\frac{1}{12}$

5 (a) $\frac{7}{20}$ (b) $\frac{1}{56}$ (c) $\frac{1}{10}$ (d) $\frac{13}{20}$
   (e) $\frac{3}{14}$ (f) $\frac{13}{18}$ (g) $\frac{5}{9}$ (h) $\frac{11}{14}$

6 (a) $3\frac{1}{2}$ (b) $3\frac{1}{3}$ (c) $7\frac{1}{2}$

Area

Essential

Angle measurer, set square

Practice booklet pages 22 to 25

A   Area of a parallelogram (p 66)

The aim is for pupils to justify for themselves the formula for the area of a parallelogram and to gain confidence in using the correct dimensions. The work provides an opportunity for pupils to practise accurate drawing.

Angle measurer, set square

◊   Drawing the parallelogram from the sketch gives an opportunity to revise work on parallel lines and constructions. The size of the obtuse angle needs to be found (110°).

◊   These points should emerge from dissecting the parallelogram.

• There are essentially two ways of dissecting the parallelogram, cutting at right angles to the longer side or at right angles to the shorter side.

• These two ways justify the two ways of measuring base and height for use in the formula.

The parallelogram has an area of about 66 cm². It can be made into a rectangle 7 cm by 9.4 cm or 10 cm by 6.6 cm.

A set square should be used as a guide when measuring perpendicular to a side.

◊   The approach above breaks down in the case of an 'overhanging' parallelogram, which will make a rectangle if cut at right angles to its longer side but not its shorter side. However, the approach in question A3 shows that the 'base × height' formula still applies when the shorter side is the base.

B Area of a triangle (p 69)

◊ The main difficulty is identifying the correct lengths to multiply by. You could ask 'What parallelogram could the triangle be half of?'
There are of course three for every triangle:

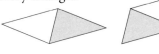

The next question is
'Which of the parallelograms do we have the measurements for?'

Again, a set square should be used as a guide when measuring perpendicular to a 'base'.

C Area of a trapezium (p 71)

◊ Draw a trapezium on the board, with the lengths of its parallel sides and the distance between them given. Pupils could work in pairs or small groups to try to find its area.

Pupils may all approach the problem in the same way, but you could get variations. Pupils who solve the problem quickly can be asked to try to find a different way of dissecting the trapezium.

◊ After discussing the approaches used, you could refer to some of the variations below.

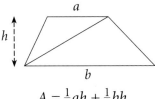

$$A = \tfrac{1}{2}ah + \tfrac{1}{2}bh$$

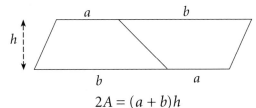

$$2A = (a + b)h$$

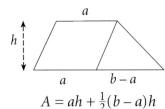

$$A = ah + \tfrac{1}{2}(b - a)h$$

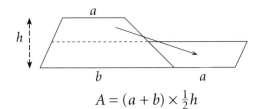

$$A = (a + b) \times \tfrac{1}{2}h$$

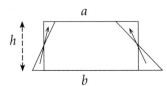

$$A = h \times \frac{(a + b)}{2}$$

Some people find it easier to think of the formula for a trapezium as
'find the mean of the lengths of the parallel sides by adding them and dividing by 2,
then multiply by the distance between them'
For this interpretation the formula can be written as

$$A = \frac{(a + b)}{2} h$$

The more complicated questions (C5 and C6) could be used to practise efficient use of a calculator's memory.

All the answers based on measurements can be expected to differ slightly from these.

Ⓐ Area of a parallelogram (p 66)

A1 About 69 cm^2

A2 About 48 cm^2

A3 P and R are equal in area,
because A + P + B = R + A + B

A4 (a) 10 cm^2 (b) 22.62 cm^2
    (c) 13.02 cm^2 (d) 29.11 cm^2

A5

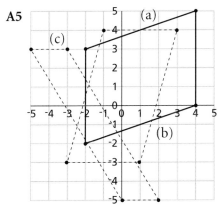

   (a) 30 square units
   (b) 28 square units
   (c) 16 square units

A6 45.72 cm^2

A7 $a = 2.4$ m $b = 1.8$ m $c = 2.5$ m
    $d = 2.1$ m $e = 3.0$ m

A8 $a = 3.0$ m $b = 5.0$ m $c = 7.0$ m

*A9 The area can be found by subtracting pieces from a square.

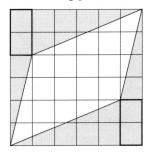

Area 18 square units

Ⓑ Area of a triangle (p 69)

B1 (a) 13.65 cm^2 (b) 11.6 cm^2
    (c) 10.2 cm^2 (d) 3.22 cm^2

B2 The pupil's answers may differ slightly because of measurement variations.

(a) Base = 5.0 cm and height = 6.0 cm or
Base = 6.1 cm and height = 4.9 cm or
Base = 7.1 cm and height = 4.2 cm
Area = 15.0 cm^2

(b) Base = 5.3 cm and height = 7.5 cm or
Base = 7.6 cm and height = 5.2 cm or
Base = 8.8 cm and height = 4.5 cm
Area = 19.8 cm^2

(c) Base = 5.0 cm and height = 7.2 cm or
Base = 7.6 cm and height = 4.7 cm or
Base = 10.3 cm and height = 3.5 cm
Area = 17.9 cm^2

B3

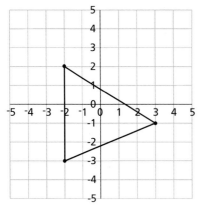

Area = 12.5 square units

B4 (a) 61.6 cm^2 (b) 43.5 cm^2
(c) 29.0 cm^2 (d) 52.2 cm^2

B5 (a) 5.5 cm   (b) 5.2 cm   (c) 6.0 cm

ℂ **Area of a trapezium** (p 71)

C1 (a) 17.0 cm^2 (b) 14.0 cm^2
(c) 31.2 cm^2

C2 17.1 cm^2

C3 The pupil's answers may differ slightly
because of measurement variations.
(a) 12.3 cm^2 (b) 23.9 cm^2

C4 22.6 m^2

C5 (a) 178.3 m^2 (to 1 d.p.)
(b) £3298 to the nearest £

C6 4411.5 m^2

C7 11.76 m^2

C8 218.2 m^2

*C9 (a) 6.4 m   (b) 10.1 m

What progress have you made? (p 74)

The pupil's answers to 1 and 4 may differ
slightly because of measurement
variations.

1 7 cm^2 2 11.7 m^2

3 6.0 m   4 4.6 cm^2

5 2.1 cm^2 6 42.75 m^2

Practice booklet

Section A (p 22)

1 (a) 15 cm^2 (b) 7.0 cm^2
(c) 22.4 cm^2 (d) 4.7 cm^2
(e) 15.75 cm^2 (f) 17.5 cm^2
(g) 25.5 cm^2 (h) 12.6 cm^2

2 (a) 12 cm   (b) 4.5 cm
(c) 4 cm   (d) 6 cm

Section B (p 23)

1 12 cm^2: A B D E G L M
18 cm^2: C H I J
24 cm^2: F K N

2 (a) 11.76 cm^2 (b) 11.96 cm^2
(c) 10.75 cm^2 (d) 11.22 cm^2
(e) 14.28 cm^2 (f) 19.74 cm^2

3 The pupil's answers may differ slightly
because of measurement variations.
(a) Triangle 1.4 cm^2, rectangle 2.8 cm^2
Total 4.2 cm^2
(b) Both parallelograms 2.8 cm^2
Total 5.6 cm^2
(c) Triangles (left to right) 3.0 cm^2,
4.0 cm^2, 2.6 cm^2
Total 9.6 cm^2
(d) Triangle 17.2 cm^2
Parallelogram 2.3 cm^2
Shaded area 14.9 cm^2
(e) Triangle 15.8 cm^2, rectangle 4.3 cm^2
Shaded area 11.5 cm^2

Section C (p 25)

1 72 cm^2: A F G
48 cm^2: B C
60 cm^2: D E

2 198.4 m^2

Review 1 (p 75)

1 $\frac{1}{25}$

2 (a)   (b)   (c)

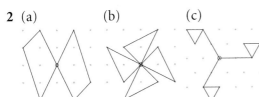

3 (a) $2x + 11$ (b) $2y + 1$ (c) $3a - 9$

   (d) $4b^2$ (e) $30c$ (f) $18d^2$

4 The perpendicular bisectors all go
through a point (the centre of the circle).

5 (a) $\frac{2}{5}, \frac{5}{12}, \frac{9}{20}$ $\left(\frac{24}{60}, \frac{25}{60}, \frac{27}{60}\right)$

   (b) (i) $\frac{5}{12}$ (ii) $1\frac{19}{30}$ (iii) $\frac{23}{24}$

6 (a) 35.36 cm^2 (b) 20.68 cm^2

7 (a) 171 cm (b) 21 cm

   (c) Overall the boys' handspans are
larger than the girls'.
The boys' handspans are less spread
out than the girls'.

8 45.9 cm^2

9 (a) The sum of the 'side' numbers is
twice the sum of the 'corner'
numbers.

   (b) Let the corner numbers be a, b, c.
The the side numbers are $a + b$,
$b + c$ and $c + a$.
The sum of the side numbers
$= a + b + b + c + c + a$
$= (a + b + c) + (a + b + c)$
which is twice $a + b + c$.

10 (a) RR, RB, RG, RY,
BR, BB, BG, BY,
GR, GB, GG, GY,
YR, YB, YG, YY

   (b) $\frac{9}{16}$

Mixed questions 1 (practice booklet p 26)

1 (a) $\frac{3}{4}$ (b) $\frac{1}{2}$ (c) $\frac{1}{2}$

   (d) $\frac{1}{4}$ (e) $\frac{1}{8}$ (f) 0

2 (a) (i) 12   (ii) 6   (iii) 3

   (b)

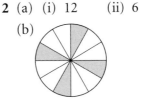

   (c) (i) 12   (ii) 6   (iii) 3

3 (a), (b), (c)

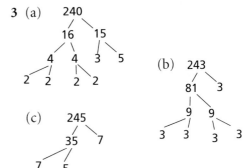

4 (a)

n	$n + 3$	$n + 6$
$n - 2$	$n + 1$	$n + 4$
$n - 4$	$n - 1$	$n + 2$

   (b) $9n + 9$ or $9(n + 1)$

5 (a)–(e)

    The pupil's constructions

   (f) Octagon

6 Median = 26 seats, range = 36 seats

7 1827 m^2

8 $\frac{13}{60}$

9 $\frac{2}{5}$

10 No. The grey triangles have a smaller
angle at the centre, so there is a smaller
probability that the arrow will land in
the grey.

Inputs and outputs

This unit introduces pupils to equivalent expressions, such as

$3(a + 6) = 3a + 18$ and $\dfrac{4a - 20}{4} = a - 5$.

p 77	**A** Input and output machines	Using number machines
p 78	**B** Shorthand rules	Using algebra to write rules
p 80	**C** Evaluating expressions	Substituting into expressions such as $2(n + 5)$, $2n + 5$
p 81	**D** Same but different!	Discussing equivalent rules such as $n \rightarrow 5(n - 3)$ and $n \rightarrow 5n - 15$
p 82	**E** Equivalent expressions	Multiplying out brackets such as $6(n + 2) = 6n + 12$
p 84	**F** Inventing puzzles	Using algebra to explain 'think of a number' puzzles

---

Essential

Sheets 148 and 149

Practice booklet pages 29 to 31

Ⓐ **Input and output machines** (p 77)

◊ You could introduce the unit to the whole class, taking questions A1 and A2 orally.

A4 These questions should get pupils thinking about looking for shorter chains. By observing the number patterns (it may be helpful to add some consecutive values to the table) they should be able to find shorter chains.

Ⓑ **Shorthand rules** (p 78)

Pupils have met the fact that we can write $a \times 3$ as $3a$. Here the notation is slightly extended, in that we write $(c + 5) \times 4$ as $4(c + 5)$.

◊ In discussing the notation, bring out the fact that any letter can be used to represent an input number, and that when multiplication is involved, the number comes first and the multiplication sign is omitted.

Mention that because $4 \times 3 = 3 \times 4$ we can (and do!) always put the number before the letter. You may want to use several examples like those in the introduction to ensure that pupils are secure with the notation. Include the fact that we can write $s \times 1$ as simply s.

B2 What pupils write here can be used diagnostically.
For example in part (a), if you see $c \rightarrow c + 12$, then it suggests the pupil is only multiplying the 4 by 3, rather than the whole of $c + 4$.

C **Evaluating expressions** (p 80)

Sheet 148

◊ Your discussion should include how to use arrow diagrams to evaluate expressions with and without brackets. You could swap the operations in the example on the page to consider $5(p - 3)$.

'Pupils enjoyed the "Cover up" game, which provided strong reinforcement.'

◊ To break the ground for the 'Cover up' game on page 81, pupils could discuss different possible rules for a single input–output pair. For example, ask pupils in pairs to find as many rules as they can that fit $2 \rightarrow 5$.

D **Same but different!** (p 81)

This teacher-led section is for pupils to begin to consider when two rules are equivalent by looking at numerical examples.
You might deal with all the examples orally.

'Spent lots of time on D1. It is very useful to refer to this exercise in later algebra work.'

◊ Emphasise that the outputs need to be the same for any input.

◊ At the end of this section ask pupils what they have discovered. Hopefully they will have seen how the expressions in each pair of results relate to each other.

E **Equivalent expressions** (p 82)

This section follows on from the teacher-led section D.
The initial discussion should include examples that involve subtraction.

Sheet 149, one between two

◊ You could also use simple areas to show equivalence.
For example

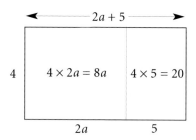

Include some examples where the number comes first in the brackets,
for example $3(2 + n) = 6 + 3n$ or $3n + 6$.

Expression snap

The game provides a way of reinforcing pupils' understanding without
too many tedious examples.

◊ You may wish to copy the sets of cards on to different coloured card. If
groups of pupils have different coloured sets, it will help them sort the
cards after each game.

F Inventing puzzles (p 84)

Pupils use the algebra they have learned to explain how 'think of a
number' type puzzles work. Some of the puzzles give a fixed number,
others end up with the starting number.

◊ Pupils have found it fun if the teacher reads out one puzzle to the class,
and then asks individuals what their answers are. Amazingly (?) they are
all the same. Pupils might then discuss in groups why this is so. One
representative from each group may then give the group's explanation to
the whole class.

Here is a way to write out an algebraic explanation.

Think of a number.	n
Multiply it by 2.	$2n$
Add on 6.	$2n + 6$
Divide by 2.	$\frac{2n + 6}{2} = n + 3$
Take off the number you first thought of.	$n + 3 - n = 3$
What is your answer?	3

Ⓐ Input and output machines (p 77)

A1 (a) $17 \xrightarrow{\times 2} \mathbf{34} \xrightarrow{-5} \mathbf{29}$

(b) $6 \xrightarrow{\times 3} \mathbf{18} \xrightarrow{-4} 14$

A2 (a) $2 \to \mathbf{2}$ (b) $6 \to \mathbf{7}$
$1 \to \mathbf{^{-}1}$ $9 \to \mathbf{8\frac{1}{2}}$
$\mathbf{8} \to 20$ $\mathbf{0} \to 4$
$\mathbf{0} \to {^{-}4}$ $3 \to 5\frac{1}{2}$

(c) $9 \to \mathbf{28}$ (d) $10 \to \mathbf{11}$
$4 \to \mathbf{18}$ $7 \to \mathbf{9.5}$
$\mathbf{5} \to 20$ $5 \to 8.5$
$\frac{1}{2} \to 11$ $72 \to 42$

A3 (a) $2 \to \mathbf{10}$ (b) $1 \to \mathbf{2}$
$8 \to \mathbf{28}$ $5 \to \mathbf{6}$
$\mathbf{10} \to 34$ $\mathbf{39} \to 40$
$\mathbf{5} \to 19$ $\mathbf{9} \to 10$

(c) $1 \to \mathbf{10}$ (d) $10 \to \mathbf{7}$
$2.1 \to \mathbf{15.5}$ $5 \to \mathbf{4.5}$
$\mathbf{7} \to 40$ $\mathbf{4} \to 4$
$\mathbf{1.1} \to 10.5$ $\mathbf{8.4} \to 6.2$

*****A4** (a) $\times 3, + 4$

(b) $+ 1$

(c) $\times 5, + 5$ or $+ 1, \times 5$

(d) $\div 2, + 2$ or $+ 4, \div 2$

Ⓑ Shorthand rules (p 78)

B1 (a) $a \xrightarrow{\times 3} \mathbf{3a} \xrightarrow{-2} \mathbf{3a - 2}$

(b) $a \to 3a - 2$

B2 (a) $c \to 3(c + 4)$ (b) $p \to 2p + 4$

B3 A is correct.

B4 (a) $a \to 5a - 3$ (b) $a \to 5(a - 3)$
(c) $w \to \frac{w + 7}{2}$ (d) $w \to \frac{w}{2} + 7$

B5 (a) $s \xrightarrow{\times 4} 4s \xrightarrow{+5} 4s + 5$

(b) $t \xrightarrow{-5} t - 5 \xrightarrow{\div 3} \frac{t - 5}{3}$

(c) $w \xrightarrow{\times 5} 5w \xrightarrow{-7} 5w - 7$

(d) $x \xrightarrow{\div 4} \frac{x}{4} \xrightarrow{-1} \frac{x}{4} - 1$

(e) $y \xrightarrow{\times 7} 7y \xrightarrow{+3} 7y + 3$

(f) $z \xrightarrow{+5} z + 5 \xrightarrow{\times 2} 2(z + 5)$

B6 $f \xrightarrow{\times 2} 2f \xrightarrow{+3} 2f + 3 \xrightarrow{\div 5} \frac{2f + 3}{5}$

The rule is $f \to \frac{2f + 3}{5}$.

B7 (a) $d \to \frac{d + 10}{5}$ (b) $s \to 3(s + 4)$

(c) $g \to \frac{g + 4}{2}$ (d) $a \to \frac{a - 10}{2}$

(e) $e \to 6e - 2$ (f) $h \to \frac{h}{4} - 2$

The pupil's explanations

Ⓒ Evaluating expressions (p 80)

C1 16

C2 (a) 8   (b) 20   (c) 60   (d) 0

C3 (a) $5 \to \mathbf{1}$ (b) $6 \to \mathbf{1\frac{1}{2}}$ (c) $10 \to \mathbf{3\frac{1}{2}}$

C4 (a) $1 \to 4$

(b) $1 \to 4$ and $2 \to 9$

(c) $1 \to 4$ and $5 \to 12$

(d) $1 \to 4$ and $5 \to 0$

(e) $2 \to 9$

(f) $4 \to 5$

C5 The pupil's three rules for $4 \to 0$

Cover up

One solution to board A:

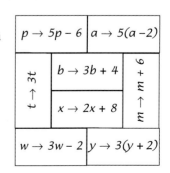

A solution to board B:

$w \to 3w - 2$		$p \to 5p - 6$	$b \to 3b + 4$
$y \to 3(y+2)$	$a \to 5(a-2)$	$m \to m + 6$	
	$t \to 3t$	$x \to 2x + 8$	

D Same but different! (p 81)

D1 (a) $+4$ (b) -6 (c) $+6$ (d) -4

E Equivalent expressions (p 82)

E1 (a) $2a + 8 = \mathbf{2}(a + 4)$

(b) $4a - 12 = 4(a - \mathbf{3})$

(c) $\dfrac{a - 6}{2} = \dfrac{a}{2} - \mathbf{3}$

(d) $5a - 20 = \mathbf{5}(a - \mathbf{4})$

(e) $\dfrac{a + \mathbf{36}}{3} = \dfrac{a}{\mathbf{3}} + 12$

(f) $\mathbf{3}a + 6 = 3(a + \mathbf{2})$

E2 $3x + 18$ and $3(x + 6)$

$3(x - 2)$ and $3x - 6$

$3x - 18$ and $3(x - 6)$

The odd one is $3x - 2$.

E3 (a) $3x - 27$　(b) $\dfrac{b}{2} + 5$

(c) $5c - 150$　(d) $4d - 8$

(e) $0.5e + 3$　(f) $8w + 4$

(g) $\dfrac{a}{5} + 4$　(h) $70m - 7$

(i) $12 + 6f$　(j) $27 + 3y$

(k) $3 + \dfrac{k}{2}$　(l) $1.5n + 30$

E4 (a) $a \to 4(3a - 2)$,　$a \to 12a - 8$;
28 when $a = 3$,　52 when $a = 5$

(b) $a \to 3(2a - 4)$,　$a \to 6a - 12$;
6 when $a = 3$,　18 when $a = 5$

(c) $a \to 2(4a - 3)$,　$a \to 8a - 6$;
18 when $a = 3$,　34 when $a = 5$

E5 (a), (b) Working leading to $6a + 10$

E6 (a) $8x + 2$　(b) $12 + 20y$

(c) $6z - 9$　(d) $30 + 15w$

(e) $16 - 4u$　(f) $v + 6$

(g) $4 + 5t$　(h) $3s - 4$

(i) $12 + 1.5r$　(j) $3 - \frac{1}{3}q$

E7 (a) $3(2a + 3) = \mathbf{6}a + 9$

(b) $4(4c + \mathbf{3}) = \mathbf{16}c + 12$

(c) $5(\mathbf{3}e + \mathbf{6}) = 15e + 30$

(d) $\mathbf{3}(2d + 7) = 6d + 21$

(e) $\dfrac{12e + \mathbf{8}}{4} = 3e + 2$

(f) $\dfrac{8g + 24}{8} = g + \mathbf{3}$

F Inventing puzzles (p 84)

F1 You always get 5.

Think of a number.	n
Multiply it by 4.	$4n$
Add on 20.	$4n + 20$
Divide by 4.	$n + 5$
Take off your first number.	5
What is your answer?	5

F2 You get the number you first thought of.

Think of a number.	n
Multiply it by 2.	$2n$
Add on 10.	$2n + 10$
Divide by 2.	$n + 5$
Take off 5.	n
What is your answer?	n

F3 (a) You always get 0.

Think of a number.	n
Add on 5.	$n + 5$
Multiply it by 4.	$4n + 20$
Subtract 20.	$4n$
Divide by 4.	n
Take off your first number.	0
What is your answer?	0

(b) You always get 2.

Think of a number.	n
Add on 3.	$n + 3$
Multiply it by 2.	$2n + 6$
Subtract 6.	$2n$
Divide by 2.	n
Add 2.	$n + 2$
Take off your first number.	2
What is your answer?	2

F4 (a)
Think of a number.	n
Subtract 2.	$n - 2$
Multiply it by 3.	$3n - 6$
Add 14.	$3n + 8$
Add your first number.	$4n + 8$
Divide by 4.	$n + 2$
Take off your first number.	2
What is your answer?	2

(b)
Think of a number.	n
Subtract 4.	$n - 4$
Multiply it by 5.	$5n - 20$
Add 20.	$5n$
Take off your first number.	$4n$
Divide by 4.	n
What is your answer?	n

F5 The missing lines could be

Multiply it by 2.
Add on 18.
Divide by 2.
Take off your first number.

F6 ? is 16.

F7
Think of a number.	n
Multiply it by 3.	$3n$
Add 15.	$3n + 15$
Divide by 3.	$n + 5$
Take off 5.	n

F8 The pupil's puzzle yielding 10

F9 The pupil's puzzle ending with start number

F10 (a) The pupil's input and outputs
(b) The outputs differ by 3.
(c) If the input number is n, we have

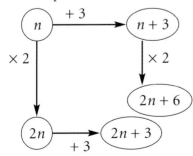

(d) The pupil's further investigations

What progress have you made? (p 86)

1 (a)
$y \xrightarrow{\times 3} 3y \xrightarrow{-6} 3y - 6$

(b)
$x \xrightarrow{+4} x + 4 \xrightarrow{\div 2} \dfrac{x+4}{2}$

2 (a) 19   (b) $12\frac{1}{2}$ (c) $1\frac{1}{2}$

3 (a) $4x - 12$ (b) $5(3s + 2)$
(c) $\dfrac{h}{2} + 6$ (d) $6y - 15$
(e) $30z + 6$ (f) $4e + 2$

4
Think of a number.	n
Multiply it by 5.	$5n$
Add 50.	$5n + 50$
Divide by 5.	$n + 10$
Take off 3.	$n + 7$
Take off your first number.	7
What is your answer?	7

Practice booklet

Sections A and B (p 29)

1 $1 \rightarrow \mathbf{4}$
$4 \rightarrow \mathbf{10}$
$\mathbf{6} \rightarrow 14$
$\mathbf{10} \rightarrow 22$

2 $2 \rightarrow \mathbf{14}$
$6 \rightarrow \mathbf{38}$
$\mathbf{3} \rightarrow 20$
$\mathbf{0} \rightarrow 2$

3 (a)
$a \xrightarrow{-3} a - 3 \xrightarrow{\times 4} \mathbf{4(a - 3)}$
$a \rightarrow 4(a - 3)$

(b)
$a \xrightarrow{\times 4} \mathbf{4a} \xrightarrow{-3} \mathbf{4a - 3}$
$a \rightarrow 4a - 3$

(c)
$e \xrightarrow{\div 2} \dfrac{e}{2} \xrightarrow{-1} \dfrac{e}{2} - 1$
$e \rightarrow \dfrac{e}{2} - 1$

(d)
$e \xrightarrow{-1} e - 1 \xrightarrow{\div 2} \dfrac{e - 1}{2}$
$e \rightarrow \dfrac{e - 1}{2}$

(e)

$g \to 3g - 6$

(f)

$h \to \dfrac{h+8}{4}$

(g)

$i \to 7(i - 12)$

(h)

$s \to 8s + 2$

4 (a)

(b)

(c)

(d)

(e)

(f)

(g)

(h)

or $\boxed{5(3 + 4c)}$

Sections C and D (p 30)

1 (a) (i) 10   (ii) 22   (iii) 2

(b) (i) 35   (ii) ⁻1   (iii) 119

(c) (i) 2   (ii) 1   (iii) 6.5

2 $3(m + 1)$ and $5(m + 1)$

3 (a) 4   (b) 1   (c) ⁻5

(d) 3.5   (e) 30   (f) ⁻4

4 (a) + 1   (b) − 4   (c) + 4   (d) + 2

Sections E and F (p 31)

1 (a) $2(x + 3) = 2x + \mathbf{6}$

(b) $\mathbf{4}(x + 4) = 4x + 16$

(c) $5x + 10 = \mathbf{5}(x + \mathbf{2})$

(d) $\dfrac{x - \mathbf{8}}{2} = \dfrac{x}{\mathbf{2}} - 4$

(e) $\dfrac{x + 12}{2} = \dfrac{x}{2} + \mathbf{6}$

(f) $\mathbf{4}(x + 6) = 4x + 24$

2 (a) $2a + 6$ (b) $5x - 5$

(c) $\dfrac{p}{4} + 2$ (d) $10p + 40$

(e) $4s + 10$ (f) $100c - 100$

(g) $\dfrac{y}{2} - 1.5$ (h) $30m - 6$

(i) $8a - 4$ (j) $2j + 1$

(k) $2k - 6$ (l) $40a - 10$

3 (a) The answer is always 2.

Think of a number.	x
Add on 3.	$x + 3$
Multiply it by 2.	$2x + 6$
Subtract 2.	$2x + 4$
Divide by 2.	$x + 2$
Subtract your first number.	2

(b) The answer is always double the starting number.

Think of a number.	x
Multiply it by 5.	$5x$
Add on 10.	$5x + 10$
Divide by 5.	$x + 2$
Double it.	$2x + 4$
Subtract 4.	$2x$

(c) The answer is the starting number.

Think of a number.	x
Subtract 1.	$x - 1$
Multiply by 3.	$3x - 3$
Add 6.	$3x + 3$
Divide by 3.	$x + 1$
Subtract 1.	x

(d) The answer is 6 times the starting number.

Think of a number.	x
Multiply by 10.	$10x$
Subtract 5.	$10x - 5$
Divide by 5.	$2x - 1$
Multiply by 3.	$6x - 3$
Add 3.	$6x$

Decimals

Practice booklet pages 32 to 34

Ⓐ **Multiplying and dividing by 10, 100, 0.1, …** (p 87)

Ⓑ **From 3 × 2 to 30 × 200, 0.3 × 0.2, …** (p 88)

Pupils use tables facts and multiplication and division by 10 to multiply without a calculator.

◊ Each line in the calculation follows from the previous one by multiplying or dividing one of the numbers by 10. After some practice, pupils may not need to use so many steps. For example, 300×0.004 could be done by multiplying 12 by 10 twice and then dividing by 10 three times. The result is the same as if it had been divided by 10 once, that is 1.2.

Ⓒ **Rounding to one significant figure** (p 90)

◊ Although we use the phrase 'significant figure', it would be better if it were 'significant place value'. The most significant place value is the highest one which is not occupied by a zero.

You can introduce the symbol ≈ to mean 'is approximately equal to' at this stage if you wish.

D Estimation (p 91)

Pupils estimate the result of a calculation by rounding numbers to 1 s.f.

◊ Estimating the result of a division is more subtle than multiplication.
For example, when estimating 476 ÷ 5.9 it is easier to think of 480 ÷ 6 than
500 ÷ 6. The division calculations in the exercise have been chosen to avoid
this kind of subtlety.

E Written multiplication (p 92)

◊ The position of the decimal point can be decided either by reference to the
rough estimate or by the method used in section B, for example:

$234 \times 6 = 1404$ leads to $23.4 \times 6 = 140.4$, and so to $23.4 \times 0.6 = 1.404$

F Dividing by a decimal (p 92)

G Written division (p 93)

◊ It helps to write the numbers with decimal points aligned, for example:

$$\frac{28.8}{0.24}$$

H Rounding to two or more significant figures (p 94)

◊ Some pupils may think that in a number like 306 584, the second
significant figure is the 6. This is an understandable mistake, seeing as the 0
signifies nothing! However, once the first significant place value has been
identified, the next place value is the second, the next the third and so on.

H1 Parts (h) to (j) bring out the point above.

Another possible cause of confusion is when zeros are significant and must
be recorded. For example, 2.976 to 3 s.f. is 3.00. This point is avoided in the
questions. Depending on how confident the pupils feel, you could
introduce it. But it might be better left for later or until it arises in the
course of a particular problem.

I Different ways of rounding (p 94)

Pupils round numbers to a given number of significant figures or a given
number of decimal places.

◊ At this stage it is probably too difficult for pupils to appreciate why there
are two ways of rounding. It has to do with relative accuracy.

If a number is expressed to, say, 2 s.f., then whatever the size of the number, the error (from the third significant figure onwards) has about the same relative size. But if two numbers are both expressed to, say 2 d.p., the errors can be of very different relative sizes. The possible error in 352.68 (i.e. 0.005) is small relative to the size of the number, but in 0.68 it is much larger.

In measurements, a result expressed to, say, 3 s.f., has the same degree of accuracy whatever the metric units used. For example, the lengths 1.64 m, 164 cm and 1640 mm are expressed to the same degree of accuracy.

Ⓐ Multiplying and dividing by 10, 100, 0.1, ... (p 87)

A1 (a) 6.5   (b) 0.3678   (c) 0.2743
   (d) 12   (e) 0.84   (f) 0.0337
   (g) 5.46   (h) 0.003 08   (i) 501
   (j) 0.000 898   (k) 0.0262   (l) 2.37

A2 (a) 2300 m   (b) 37 g
   (c) 0.0173 km   (d) 2.08 litres
   (e) 0.009 77 m   (f) 0.023 04 km

A3 (a) 0.0865   (b) 367.8   (c) 2.243
   (d) 0.0569   (e) 0.0064   (f) 287
   (g) 0.536   (h) 70 800

A4 (a) 0.0023 m   (b) 7.5 g   (c) 320 cm
   (d) 50.3 ml   (e) 0.0305 m   (f) 95.5 mm

A5 (a) 0.01   (b) 100   (c) 0.1
   (d) 1000   (e) 0.01   (f) 1000

A6 $23\,750 \times 0.001 = 23.75$
$237.5 \times 0.01 = 2.375$
$0.002\,375 \div 0.1 = 0.023\,75$

Ⓑ From 3 × 2 to 30 × 200, 0.3 × 0.2, ... (p 88)

B1 (a) 6000   (b) 60 000   (c) 6
   (d) 60   (e) 0.06   (f) 2000
   (g) 2 000 000   (h) 200   (i) 2
   (j) 20

B2 (a) 0.18   (b) 0.12   (c) 180
   (d) 30   (e) 1.6

B3 (a) $5 \times 6, 0.5 \times 60, 50 \times 0.6,$
     $0.05 \times 600, 500 \times 0.06$
   (b) $5 \times 0.6, 0.5 \times 6, 50 \times 0.06, 0.05 \times 60$
   (c) $5 \times 60, 0.5 \times 600, 500 \times 0.6, 50 \times 6$
   (d) $0.5 \times 50, 0.05 \times 500$
   (e) $0.05 \times 50, 0.5 \times 5$ (f) 0.06×6

B4 (a) 32 m   (b) 500   (c) 50
   (d) 500   (e) 120

B5 (a) 6   (b) 1.6
   (c) 23.2 ($-10, \times 0.4, \times 0.2, +20$ or
     $-10, \times 0.2, \times 0.4, +20$)

B6 41

B7 $50 \times 40 = 2000, \quad 0.5 \times 400 = 200,$
$0.4 \times 50 = 20$

B8 $0.3 \times 4 = 1.2, \quad 0.2 \times 0.3 = 0.06,$
$30 \times 0.2 = 6, \quad 0.4 \times 30 = 12$

Ⓒ Rounding to one significant figure (p 90)

C1 The figure 4 is the most significant. 4000

C2 (a) 4000   (b) 8000   (c) 20 000
   (d) 800   (e) 800 000   (f) 40 000

C3 (a) 0.3   (b) 0.05
   (c) 0.003   (d) 0.0006

C4 (a) 0.07   (b) 100 000
   (c) 90 000   (d) 0.007

C5 (a) 50   (b) 2   (c) 0.0008   (d) 0.08

D Estimation (p 91)

D1 (a) $30 \times 80 = 2400$

(b) $400 \times 30 = 12\,000$

(c) $400 \times 200 = 80\,000$

(d) $200 \times 4000 = 800\,000$

(e) $5000 \times 400 = 2\,000\,000$

(f) $2000 \times 300 = 600\,000$

D2 (a) $70 \times 9 = 630$ (b) $0.5 \times 2 = 1$

(c) $0.02 \times 50 = 1$ (d) $0.8 \times 0.2 = 0.16$

(e) $6 \times 0.02 = 0.12$

D3 (a) 1.0404 (b) 14.8732

(c) 0.096\,901 (d) 0.309\,491\,6

D4 $3 \times 10 \times 5000 = 150\,000\,\text{cm}$

D5 $20 \times 7 \times 3 = 420\,\text{m}^3$

D6 (a) $\dfrac{800 \times 0.1}{2} = 40$ (b) $\dfrac{0.5 \times 60}{10} = 3$

(c) $\dfrac{0.08 \times 20}{4} = 0.4$ (d) $\dfrac{200}{4 \times 5} = 10$

(e) $\dfrac{6000}{60 \times 0.5} = 200$ (f) $\dfrac{900}{0.9 \times 50} = 20$

E Written multiplication (p 92)

E1 (a) 33.6 (b) 1.89 (c) 20.8

(d) 0.078 (e) 0.552 (f) 440

(g) 0.0104 (h) 2.349

E2 (a) 12.48 (b) 12.48 (c) 124.8

(d) 0.1248

E3 (a) 53.08 (b) 0.5308 (c) 53.08

(d) 0.5308

F Dividing by a decimal (p 92)

F1 (a) 20 (b) 60 (c) 60

(d) 30 (e) 8

F2 (a) 7 (b) 0.4 (c) 0.6

(d) 0.2 (e) 300

F3 (a) 40 (b) 70 (c) 200 (d) 40

(e) 200 (f) 2 (g) 80 (h) 5

(i) 50 (j) 400

F4 $0.004 + 0.02 = 0.024$, $\quad 0.08 \times 3 = 0.24$,

$1.6 \div 0.04 = 40$, $\quad 0.5 - 0.02 = 0.48$

G Written division (p 93)

G1 (a) 0.805 (b) 39.6 (c) 44.8

(d) 0.0072 (e) 160 (f) 735

(g) 56 (h) 0.104 (i) 0.675

(j) 855

G2 (a) 6.03 (b) 603 (c) 0.603

(d) 0.000\,603 (e) 6030

G3 (a) 346 (b) 346 (c) 0.003\,46

(d) 3460 (e) 34.6

G4 (a) 6.025 (b) 6.025 (c) 0.006\,025

(d) 0.060\,25 (e) 602\,000

H Rounding to two or more significant figures (p 94)

H1 (a) 67\,000 (b) 0.067 (c) 0.15

(d) 460 (e) 0.0037 (f) 0.73

(g) 79 (h) 900 (i) 51\,000

(j) 1000

H2 (a) 78\,300 (b) 0.173 (c) 2000

(d) 3710 (e) 0.005\,57 (f) 0.001\,54

(g) 1280 (h) 903 (i) 851

(j) 49\,900

H3 (a) 600 (b) 570 (c) 575 (d) 574.6

I Different ways of rounding (p 94)

I1 (a) 5.33 (b) 5.3

I2 (a) 3600 (b) 4000

I3 (a) 6.6 (b) 6.52 (c) 0.008

(d) 0.009 (e) 6.5 (f) 0.007\,82

I4 (a) 3460 (b) 3000 (c) 46\,700

(d) 47\,000 (e) 7.44 (f) 7.4

I5 (a) 20 (b) 20.04 (c) 0.006

(d) 0.005\,89 (e) 5.098 (f) 5.10

What progress have you made? (p 95)

1 (a) 12 (b) 0.08 (c) 28

2 (a) 80 000 (b) 0.027 (c) 380 (d) 7.40

3 (a) $70 \times 0.5 = 35$ (b) $0.04 \times 300 = 12$

(c) $\dfrac{800 \times 0.2}{40} = 4$

4 (a) 2.482 (b) 0.0952

5 (a) 300 (b) 90

6 (a) 5.6 (b) 27

Practice booklet

Section A (p 32)

1 (a) 0.052 (b) 5.2
(c) 0.0634 (d) 634

2 (a) 100 (b) 0.01 (c) 1000
(d) 0.01 (e) 0.1 (f) 0.01

Section B (p 32)

1 (a) 15 (b) 1500 (c) 15 (d) 150

2 (a) 2800 (b) 16 (c) 360 000
(d) 600 (e) 0.2 (f) 180

3 (a) 300×0.4; 30×4; 3×40; 0.3×400
(b) 300×0.3; 30×3
(c) 30×0.04; 3×0.4; 0.3×4; 0.03×40
(d) 400×0.04; 40×0.4
(e) 3×0.04; 0.3×0.4; 0.03×4

Sections C and D (p 32)

1 (a) 0.5 (b) 7000 (c) 20
(d) 50 000 (e) 9 (f) 0.7
(g) 0.02 (h) 0.0008

2 (a) is given
(b) $300 \times 50 = 15 000$
(c) $600 \times 300 = 180 000$
(d) $5000 \times 400 = 2 000 000$
(e) $50 \times 10 = 500$

(f) $0.3 \times 0.4 = 0.12$
(g) $8 \times 90 = 720$
(h) $6000 \times 0.8 = 4800$
(i) $400 \times 0.008 = 3.2$
(j) $0.03 \times 30 = 0.9$

3 3000×5 litres $= 15 000$ litres

4 $40 \times 80\text{p} = 3200\text{p} = £32$

Section E (p 33)

1 (a) 16.17 (b) 0.111
(c) 180 (d) 0.0228

2 (a) 4.76 (b) 47.6
(c) 0.0476 (d) 476

Sections F and G (p 33)

1 (a) 7 (b) 8 (c) 90 (d) 3000
(e) 300

2 (a) 2.4 (b) 4800 (c) 0.0148
(d) 1.28 (e) 2500

Section H (p 34)

1 (a) 48 000 (b) 0.042
(c) 9.1 (d) 0.0037

2 (a) 871 (b) 0.0419
(c) 27.8 (d) 310

3 (a) 0.004 81 (b) 9.0
(c) 4 306 000 (d) 421

Section I (p 34)

1 (a) 67 800 (b) 68 000
(c) 5.4 (d) 5

2 (a) 0.04 (b) 0.040
(c) 8710 (d) 8712.5

*3 (a) 6683, 6731, 6702
(b) 6650 (c) 6749

*4 (a) £27 800, £28 428
(b) £27 500 (c) £28 499.99

⑬ Investigations

Investigative and problem-solving work are best integrated into the development of mathematical concepts and skills. However, focusing on investigative work, as here, allows important skills of report writing to be developed. It is not intended that all the investigations should be done together or in the order given.

Optional Counters, tiles or pieces of paper (for B1) Square dotty paper (for B5)

🄰 Crossing points (p 96)

Discussion of the report is intended to highlight some important processes, for example, specialising, tabulating, generalising, predicting, checking, explaining, drawing conclusions. It is not intended to suggest there is one 'correct' way to approach an investigation and write up the findings.

◊ If pupils find it difficult to follow Chris's written work, try to involve them actively. They could read through the first half of page 97 (the 1, 2 and 3 line results) and then try with 4 lines.
Emphasise that the lines should be drawn as long as possible so that all crossings are shown.

Ask pupils to find as many different numbers of crossings as possible with 4 lines; 0, 1, 3, 4, 5 and 6 are possible:

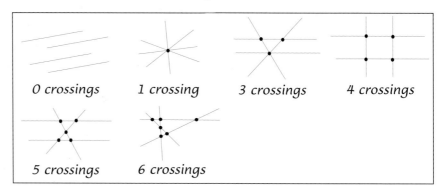

| 0 crossings | 1 crossing | 3 crossings | 4 crossings |

| 5 crossings | 6 crossings |

Ask how we can be sure that 2 crossings cannot be achieved or that 6 is the maximum number. Emphasise that pupils should try to explain their findings wherever possible.

◊ Now discuss how the investigation could proceed. Looking at the maximum number of crossing points is one choice (and it's the one made by the pupil in the write-up). Look at the results in the table. Pupils could try to spot a pattern before turning to page 99.

◊ Look at the table on page 99 and ask pupils if they can explain why the numbers of crossing points go up in the way they do. This is easier when pupils have found these results out for themselves. To get the maximum number of crossing points each additional line needs to cross all the lines in the diagram (except itself). Chris does not try to explain this in her report but without it she cannot be sure that the sequence of numbers continues in the way she describes. Point out that 'predict and check' may help to confirm results but is not foolproof (suppose there were actually 16 crossing points for 6 lines and your 'pattern' stopped you looking any further than 15).

◊ You could ask what the maximum number of crossing points would be with, say, 100 lines. Using Chris's method, this would take some time. With n lines, each line crosses $n - 1$ lines to produce $n - 1$ crossing points. However, $n(n - 1)$ counts each crossing point twice so the number of crossing points is $\frac{n(n - 1)}{2}$.

◊ As a further investigation, pupils can look at the maximum number of closed regions obtained. An interesting result is that you get the same set of numbers but with 0 included: 0, 1, 3, 6, 10, 15, …

Ⓑ Some ideas (p 100)

B1 Round table (p 100)

> Optional: Counters, tiles or pieces of paper may be useful to represent the people round the table.

◊ Pupils may find it helpful to do the investigation by moving counters or tiles (labelled A to E) round a drawing of a table. There is only one other arrangement:

Ask pupils to consider how they know there are no other arrangements. They may realise that every person has sat next to every other person so no other arrangements are possible. Ask pupils if that is what they expected – often pupils think there will be more possibilities.

◊ In one school, the investigation proved easier when the table was 'unrolled' into a straight line (remembering that the two end people are in fact sitting next to each other). For five people the two different arrangements are

ABCDE   ADBEC

◊ The results for 3 to 10 people are

Number of people	Number of arrangements
3	1
4	1
5	2
6	2
7	3
8	3
9	4
10	4

For an even number of people the rule is $a = \frac{p-2}{2}$

and for an odd number of people the rule is $a = \frac{p-1}{2}$

where a is the number of arrangements
and p is the number of people.

Since each person has two neighbours, the number of arrangements has to be the number of complete pairs of people that can sit next to an individual. These formulas give the number of complete pairs.

B2 Nine lines (p 100)

◊ Clarify that the grids of squares have to be drawn with straight lines either parallel to each other or at right angles.

As in *Crossing points* lines should be drawn as long as possible so that all possible squares are counted. For example, diagrams such as this are not considered valid.

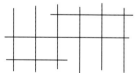

◊ With 9 lines 0, 6, 10 and 12 squares are possible:

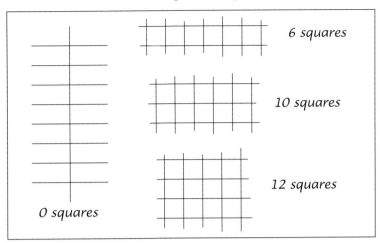

9 parallel lines will also produce 0 squares.

◊ All possible numbers of squares for 4 to 12 lines are

Number of lines	Numbers of squares				
4	0	1			
5	0	2			
6	0	3	4		
7	0	4	6		
8	0	5	8	9	
9	0	6	10	12	
10	0	7	12	15	16
11	0	8	14	18	20
12	0	9	16	21	24	25

Pupils who produce a full set of results as in the table may make various observations such as:

• The minimum is always 0 and this can be achieved by a set of parallel lines.

• The next possible number of squares is $n - 3$ for n lines. These numbers go up by 1 each time.

◊ Encourage pupils to follow their own lines of investigation. For example, they could consider the maximum number of squares possible each time.

Encourage pupils to describe how the lines should be arranged to give the maximum number of squares by asking questions such as 'How would you arrange 100 lines to achieve the maximum number of squares?' and 'What about 99 lines?'

Formulas are

even numbers of lines: $s = (\frac{n}{2} - 1)^2$

odd numbers of lines: $s = (\frac{n}{2} - \frac{1}{2})(\frac{n}{2} - \frac{3}{2})$

Again, very few pupils are likely to express their conclusions algebraically at this stage.

B3 Cutting a cake (p 101)

◊ Emphasise that each cut must go from one side of the cake to another.
For example, these cuts are not valid.

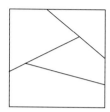

◊ Pupils could investigate the minimum and maximum number of pieces. The results are

Number of cuts	Minimum number of pieces	Maximum number of pieces
0	1	1
1	2	2
2	3	4
3	4	7
4	5	11
5	6	16
6	7	22
7	8	29
8	9	37

◊ Pupils may comment that:
- The minimum number of pieces goes up by 1 each time.
- The minimum number of pieces is always 1 more than the number of cuts (or, with n cuts, the minimum number of pieces is $n + 1$).
- The minimum number of pieces can be achieved by making a set of parallel cuts.
- The maximum number of pieces goes up by 1, then 2, then 3 and so on.

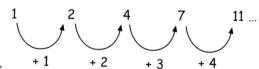

- To achieve the maximum number of pieces, take the previous diagram and make a cut that crosses each of the previous cuts.
- The sequence of numbers for the maximum number of pieces for one or more cuts (2, 4, 7, 11, 16, 22, …) can be found by adding 1 to each of the numbers in the sequence for the maximum number of crossing points in 'Crossing points' (1, 3, 6, 10, 15, 21, …).

◊ Pupils may correctly predict the maximum number of pieces for various numbers of cuts but find it difficult to produce the corresponding diagrams. Using larger squares may help.

◊ For any number of cuts all numbers of pieces between the minimum and maximum can be achieved. Each diagram can be found by altering the previous one. For example, with four cuts:

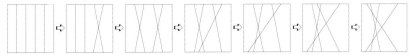

5 pieces   6 pieces   7 pieces   8 pieces   9 pieces   10 pieces   11 pieces

◊ With n cuts, the maximum number of pieces is $\frac{n(n-1)}{2} + 1$.

B4   Matchstick networks (p 101)

This investigation involves three variables.

◊ The 'standard' way to approach a three-variable problem is to keep one variable constant and investigate the relationship between the other two. However in this case the relationship between the three variables is easy enough for pupils to spot from a table, for example:

Enclosed spaces (E)	Nodes (N)	Matches (M)
0	8	7
1	8	8
2	9	10
2	7	8
4	6	4
3	8	10

The rule is $E + N - 1 = M$

◊ If the rule is not found in this way, suggest that pupils first investigate networks with no enclosed spaces ('trees'), to find the rule $N - 1 = M$. Then look at one enclosed space, two enclosed spaces, and so on.

◊ The explanation of the rule is quite difficult at this stage, but you can encourage pupils to explain what can happen when one more match is added to a network

M up by 1
N up by 1
E the same

M up by 1
N the same
E up by 1

M up by 1
N the same
E up by 1

M up by 1
N up by 1
E the same

◊ The full explanation is as follows.
For the simplest network ⟶ $E = 0, N = 2, M = 1$
so $E + N - M = 1$ for this network.

Whenever a new match is added to the network, $E + N - M$ does not change (see the four cases above). So $E + N - M$ is always 1.

B5 Turn, turn, turn (p 102)

This is a very rich activity with many possibilities for extension. Encourage pupils to follow their own lines of enquiry but some may need help in formulating questions.

Optional: Square dotty paper

◊ Many have found it beneficial for pupils to walk through the instructions to draw a turning track, emphasising the 90° turn each time.

◊ Encourage pupils to investigate questions such as:

• Do you always get back to your starting point?
   If you do, how many times do you repeat the instructions?

• What difference does it make if you turn left instead of right each time?

• What shape are the tracks?
   Do you get different types of tracks with different sets of numbers?

• Why do some tracks have 'holes'
   while others have 'overlaps'?

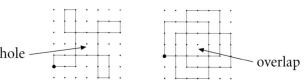

Can you predict whether or not a track will have a hole or an overlap from the set of three numbers? Can you predict the size of the hole?

• What will happen with sets of consecutive numbers?

• What if you look at sets of numbers where the first two are always the same?

• What happens if you change the order of the numbers?
   Will the track for 1, 2, 4 look like the track for 4, 1, 2 for example?

• What if two or more of the numbers are the same?

◊ Pupils may notice these facts.

• It doesn't matter what order the numbers are in. You always produce the same track although it may be rotated or reflected.

• Turning left produces a reflection of the turning right track.

• If the numbers are all the same, you get a square track.

• If two of the numbers are the same, you get a cross shape.

• All tracks made with three numbers have rotation symmetry.

- If the two smallest numbers add up to the largest, you get a track with no hole or overlap.

  For example,

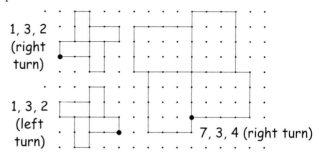

1, 3, 2 (right turn)

1, 3, 2 (left turn)

7, 3, 4 (right turn)

- If the sum of the two smallest numbers is greater than the largest number, you get a 'windmill' shape with an overlap.

  For example,

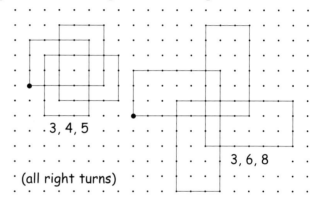

3, 4, 5

3, 6, 8

(all right turns)

- If the sum of the two smallest numbers is smaller than the largest number, you get a windmill shape with a hole.

  For example,

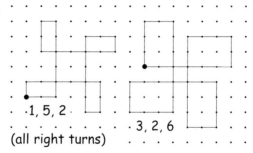

1, 5, 2

3, 2, 6

(all right turns)

The size of the hole is the largest number minus the sum of the two smallest numbers.

◊ Investigating longer sets of numbers, pupils will find that four numbers produce infinite 'spiral' tracks and five numbers produce 'closed' tracks with rotation symmetry.

For example,

2, 1, 4, 3   (all right turns)

1, 3, 4, 2, 1

.2, 3, 5, 1, 4

6, 2, 1, 3

1, 7, 2, 3, 5

◊ Pupils can investigate what happens if you change the turning angle. For example, you could use a 60° right turn and investigate on triangular spotty paper.

◊ Pupils can use LOGO to draw their turning tracks.

⑭ Parallel lines

> **Essential**
> Set square
> **Practice booklet** pages 35 to 37

Ⓐ Using parallel lines (p 103)

> Set square

◊ Pupils are likely to have met the word 'parallel' before. You could ask a pupil to draw a pair of parallel lines on the board and then ask the rest of the class to say what it is that makes the lines parallel.

◊ We are restricting our use of 'parallel' to straight lines, though the word is used in everyday speech to refer to curved lines (railway lines, lines of latitude) that never meet.

Ⓑ Parallel lines and angles (p 106)

◊ The pencil idea can be used on an OHP transparency of the diagram for question B3.

Other acceptable ways to justify corresponding and alternate angles may come up in discussion. For example, you can think of angle x as a 'solid thing' sliding along a line to become a corresponding angle; corresponding angles together with vertically opposite angles can be used to justify alternate angles.

The fact that x and y add up to 180° (that is, are supplementary) can also emerge from the B3 diagram.

Ⓒ Related angles (p 108)

◊ In one school, long rulers were placed on the floor like this.

Pupils were asked to stand on a vertically opposite/corresponding/alternate angle to the one given.

D Explaining how you work out angles (p 110)

◊ It is important for pupils to be able to explain the steps of their reasoning, even though these may appear obvious.

It is generally easier, in cases where only angles are involved, to use the method on page 110. This may involve labelling angles other than those given, hence the need to show the new labels on a copy of the diagram.

Pupils may need help on page 112 with using letters for points and three letters to label angles.

A Using parallel lines (p 103)

A1 They have a different slope: *a* goes 1 square up for every 2 along while *b* goes 1 square up for every 3 along.

A2 (a) Yes, they both go 2 squares down for every 3 along.

(b) No, *s* is steeper than *r*: *s* goes 4 down for every 2 along, but *r* goes less than 4 down for 2 along.

A3 *a*, *b*, *h* and *j*   *c*, *i* and *k*   *d*, *f* and *g*   *e* and *l*

A4

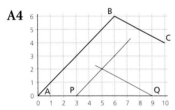

The diamond is at (5, 2).

A5

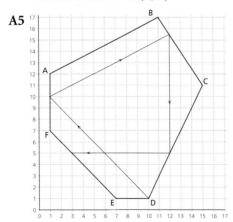

The treasure is buried at (3, 5).

A6 The two lengths measured on each transversal are the same.
If you mark points 12 cm apart and 4 cm apart, the lengths measured on each transversal are in the ratio 2 : 1.
In general, if the distances between the dots on the parallel lines are in the ratio *n* : 1, the lengths measured on each transversal are in the ratio *n* − 1 : 1.

B Parallel lines and angles (p 106)

B1 They are parallel (they go in the same direction).

B2 They are parallel (they go in the same direction).

B3 The pupil's sketch

B4 (a) $a = 130°$, $b = 50°$, $c = 130°$, $d = 50°$

(b) $e = 110°$, $f = 70°$, $g = 110°$, $h = 70°$

(c) $i = 108°$, $j = 72°$, $k = 108°$, $l = 72°$

B5 $a = 42°$, $b = 138°$
$c = 85°$, $d = 95°$
$e = 37°$, $f = 143°$

ⓒ Related angles (p 108)

C1 (a) Angles a and b are **corresponding** angles.

(b) Angles c and d are **alternate** angles.
Angles c and e are **corresponding** angles.
Angles d and e are **vertically opposite** angles.

(c) Angles f and g are **alternate** angles.
Angles f and h are **supplementary** angles.
Angles g and h are **supplementary** angles.
Angles g and i are **vertically opposite** angles.

C2 (a) 65° corresponding angles

(b) 55° alternate angles

(c) 112° vertically opposite angles

(d) 75° supplementary angles

(e) 106° supplementary angles

(f) 120° alternate angles

ⓓ Explaining how you work out angles (p 110)

Reasons need to be given for steps of working.

D1 (a) $x = 75°$ (b) $y = 95°$, $z = 65°$

(c) $v = 59°$, $w = 138°$

D2 (a) $x = 94°$, $v = 86°$

(b) $u = 76°$, $v = 99°$

(c) $r = 72°$

D3 (a) $x = 112°$, $y = 75°$

(b) $u = 132°$, $v = 56°$

(c) $p = 47°$, $q = 63°$

D4 (a) $x = 40°$ (b) $y = 110°$

(c) $z = 24°$ (d) $u = 70°$, $v = 60°$

(e) $w = 95°$ (f) $g = 105°$

D5

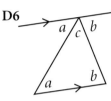

x : corresponding angles
y : alternate angles
$x + y + z = 180°$
(angles on a straight line)

The angles of a triangle add up to 180°.

D6

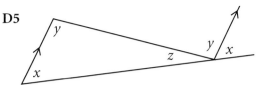

a : alternate angles
b : alternate angles
$a + c + b = 180°$
(angles on a straight line)

D7 (a) $p = 43°$, $q = 137°$

(b) $r = 84°$, $s = 142°$

(c) $t = 70°$, $u = 110°$, $v = 70°$

(d) $w = 38°$, $x = 104°$

(e) $a = 47°$, $b = 23°$, $c = 110°$

D8 Here are some possible explanations.

(a) Angle HGB + angle DGH = 180°
(angles on a straight line)
So **angle HGB = 63°**

Angle ADC = angle DGF
(corresponding)
= 180° − 117° = **63°**

(b) Angle IJM = angle OMN = 63°
(corresponding angles)

Angle IJL + angle IJM = 180°
(angles on a straight line)
So **angle IJL = 117°**

(c) Angle UTQ = 180° − 109° = 71°

Angle PQS = angle UTQ = **71°**
(corresponding angles)

D9 Lines AD and BC are parallel (angles

DAC and ACB are alternate angles).

Lines DE and CF are parallel (angles EDF and DFC are alternate angles.)

Lines AC and EF are parallel (angles ACF and CFE are supplementary angles between parallels).

What progress have you made? (p 113)

1 The pupil's drawing

2 They are not parallel.
On the top line you go up 2 when you go along 3. On the other line you go up more than 2 when you go along 3.

3 $a = 60°$ $b = 95°$ $c = 35°$

4 Angle DEB + angle BEF = 180°
(supplementary angles on a straight line)
So **angle DEB = 80°**

Angle EFC = angle DEB = **80°**
(corresponding angles)

Angle BCF = angle GBC = **75°**
(alternate angles)

Practice booklet

Section A (p 35)

1 (a) The pupil's drawing
(b) The pupil's drawing
(c) The 'diamond' shape has its sides all the same length.
It has two pairs of supplementary angles. The diagonals cut one another in half, crossing at right angles.

2 (3, 4)

3 (6, 4)

Sections C and D (p 36)

1 Angles a and d are **vertically opposite**.
Angles e and a are **corresponding**.
Angles c and f are **alternate**.
Angles f and d are **supplementary**.
Angles h and d are **corresponding**.
Angles f and e are **supplementary**.
Angles f and g are **vertically opposite**.
Angles b and f are **corresponding**.

2 (a) $p = 115°, q = 108°$
(b) $r = 50°, s = 45°$
(c) $t = 48°, u = 75°$
(d) $v = 62°, w = 68°$

3 (a) $r = 29°, s = 151°, t = 47°, u = 47°$
(b) $v = 35°, w = 145°, x = 37°, y = 143°,$

4 (a) $p = 63°, q = 45°$
(b) $r = 70°, s = 62°$
(c) $t = 62°, u = 30°$
(d) $v = 117°$
(e) $w = 38°$
(f) $x = 46°$

5 (a) Angle EDG = 100° (Angles EDG and CDG are supplementary angles on a straight line.)
Angle FEH = 100° (Angles EDG and FEH are corresponding angles.)
(b) Angle LMJ = 110° (Angles OLM and LMJ are alternate angles.)
Angle JMN = 70° (Angles LMJ and JMN are supplementary angles on a straight line.)
(c) Angle RSX = 85° (Angles VST and RSX are vertically opposite angles.)
Angle QRW = 85° (Angles RSX and QRW are corresponding angles.)

⑮ Percentage

This unit covers changing a percentage to a decimal, calculating a percentage of a quantity and expressing one quantity as a percentage of another. The latter leads on to drawing pie charts using a circular percentage scale.

Essential

Pie chart scale

Practice booklet pages 38 to 42

Ⓐ Understanding percentages (p 114)

◊ Proportions are given in a variety of ways. From class or group discussion should emerge the need for a 'common currency' for expressing and comparing proportions.

◊ French cheeses may be labelled '40% matière grasse', for example; this means 40% of the dry matter is fat (i.e. the water content of the cheese is discounted).

◊ The order is:
Mascarpone 45%, Blue Stilton 36%, Red Leicester $33\frac{1}{3}$%, Danish Blue 28%, Edam 25%, Camembert 21%, Cottage Cheese 2%, Quark 0.2%

Ⓑ Percentages in your head (p 116)

Mental work with percentages can be returned to frequently in oral sessions.

Ⓒ Percentages and decimals (p 117)

Ⓓ Calculating a percentage of a quantity (p 118)

It is important that pupils should feel confident about their method of working out a percentage of a quantity. The 'one-step' method is more sophisticated but ultimately better because it easily extends to a succession of percentage changes.

Ⓔ Changing fractions to decimals (p 119)

You could point out that the division symbol ÷ is itself a fraction line with blanks above and below for numbers.

F One number as a percentage of another (p 120)

The approach used here depends on conversion from decimal to percentage.

G Drawing pie charts (p 122)

> Pie chart scale

◊ The pie charts shown are not labelled with the percentages, in order to give practice in measuring. However, it is a good practice to include the percentages.

◊ Rounding often leads to percentages which add up to slightly more or less than 100%. For this reason, when drawing pie charts it is often better to work to the nearest 0.1% as the small excess or deficit can be safely ignored.

H Problems involving percentages (p 124)

A Understanding percentages (p 114)

A1 (a) About 27% (b) Water
(c) About 21%

A2 (a) C, D, F (b) G, H
(c) A, E (d) B, I

A3 (a) 20%–30% (b) 85%–95%
(c) 45%–55% (d) 55%–65%
(e) 3%–8%

B Percentages in your head (p 116)

B1 (a) $\frac{1}{4}$ (b) $\frac{3}{4}$ (c) $\frac{1}{10}$
(d) $\frac{9}{10}$ (e) $\frac{1}{5}$

B2 (a) 50% (b) 10% (c) 25%
(d) 75% (e) $33\frac{1}{3}$%, 25%

B3 (a) £15 (b) £42 (c) £17.50

B4 (a) £10 (b) £21 (c) £17.50

B5 (a) To find 10%, you can divide by 10.
(b) To find 5%, you can divide by 10, then halve.

B6 (a) 1p (b) 3p (c) 37p

B7 (a) £10.50 (b) £38.50

C Percentages and decimals (p 117)

C1 (a) 0.5 (b) 0.25 (c) 0.65 (d) 0.78
(e) 0.1 (f) 0.01 (g) 0.04 (h) 0.4

C2

$\frac{45}{100}$	=	0.45	=	**45%**
$\frac{57}{100}$	=	0.57	=	**57%**
$\frac{5}{100}$	=	0.05	=	**5%**
$\frac{63}{100}$	=	0.63	=	63%
$\frac{7}{100}$	=	0.07	=	**7%**

C3 (a) 30% (b) 80% (c) 83% (d) 3%

C4 1%, 0.1, $\frac{12}{100}$, 15%, 0.25, 0.3, $\frac{45}{100}$

D Calculating a percentage of a quantity (p 118)

D1 (a) 162 g (b) 256.2 g (c) 66.7 g
(d) 93.8 g (e) 26.6 g (f) 35.2 g
(g) 201.6 g (h) 52.8 g

D2 (a) 2.25 g   (b) 6.72 g
     (c) 3.6 g   (d) 1.68 g

D3 0.6 g

D4 He is right for 10%, but dividing by 5 gives 20%, not 5%.

D5 (a) Sugar 19.95 g, fat 10.5 g, protein 2.8 g
     (b) Sugar 85.5 g, fat 45 g, protein 12 g
     (c) Sugar 285 g, fat 150 g, protein 40 g

E Changing fractions to decimals (p 119)

E1 (a) 0.25   (b) 0.125   (c) 0.05
     (d) 0.8   (e) 0.375   (f) 0.875
     (g) 0.28   (h) 0.15   (i) 0.22
     (j) 0.9375

E2 $\frac{29}{50}$ (0.58), $\frac{3}{5}$ (0.6), $\frac{5}{8}$ (0.625), $\frac{13}{20}$ (0.65)

E3 (a) 0.14 (b) 0.57 (c) 0.11 (d) 0.56
     (e) 0.64 (f) 0.27 (g) 0.08 (h) 0.38
     (i) 0.41 (j) 0.87

F One number as a percentage of another (p 120)

F1 71% (to the nearest 1%)

F2 (a) 60%   (b) 35%   (c) 87.5%
     (d) about 33%   (e) about 67%

F3 (a) 29%   (b) 78%   (c) 23%
     (d) 41%   (e) 5%

F4 (a) $\frac{1}{4}$ (b) 25%   (c) 19%

F5 30%

F6 Tina's method is correct.

F7 (a) 16%
     (b) In a class of 30 it would be about 5 people.

F8 12.4 × 0.41 = 5.084

F9 (a) 64%   (b) 54%   (c) 76%
     (d) 35%   (e) 74%   (f) 69%

F10 A 14.0% fat   B 15.5% fat
B has higher percentage of fat.

G Drawing pie charts (p 122)

G1 These features (amongst others) may be noticed:
Cheese spread has higher proportions of water and carbohydrate, but lower proportions of fat and protein.

G2 (a) 27%   (b) 23%   (c) 46%   (d) 53%

G3

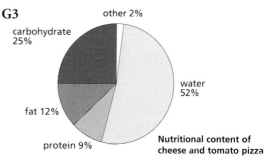

Nutritional content of cheese and tomato pizza

G4 (a) Meat, fish and eggs   (b) 9%
     (c) Clio is wrong. The chart shows money spent, not the amounts eaten.

G5

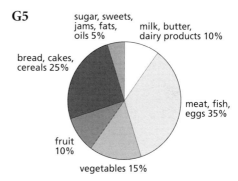

G6 (a) 34%
     (b) Foreign news 22%, sport 16%, entertainment 9%, finance 19%

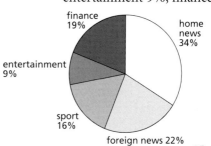

G7

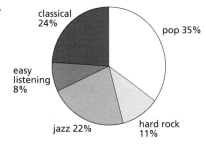

classical 24%
pop 35%
easy listening 8%
jazz 22%
hard rock 11%

Ⓗ Problems involving percentages (p 124)

H1 287.5 ml

H2 40 g

H3 (a) No. If he gets 30 g protein he will get 60 g fat.

(b) With cheese B, if he gets 30 g protein he will get 22.5 g fat. With cheese C, if he gets 30 g protein he will get 15 g fat. So cheese C will suit him.

H4 (a) 50%   (b) 37.5%

(c) 91.7%   (d) 58.3%

H5 12.1%

H6 (a)

of	20	10	50
5%	1	0.5	2.5
1%	0.2	0.1	0.5
8%	1.6	0.8	4

(b)

of	50	30	80
15%	7.5	4.5	12
90%	45	27	72
1%	0.5	0.3	0.8

What progress have you made? (p 125)

1 (a) C   (b) A   (c) E

2 (a) 0.5 (b) 0.45 (c) 0.04 (d) 0.07

---

3 (a) £5   (b) 2 kg   (c) 9 kg   (d) £2

4 (a) 68.4 g   (b) 13 g

5 73%

6 21 out of 25 (84%) is better than 30 out of 37 (81%).

7

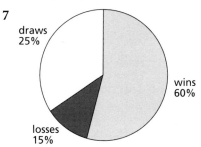

draws 25%
wins 60%
losses 15%

Practice booklet

Section A (p 38)

1 Cadbury's Double Decker:
protein 6%, fat 22%,
carbohydrate 72%

Sainsbury's Creamy White:
protein 7%, fat 35%,
carbohydrate 58%

2 (a) 30% (b) 10% (c) 70% (d) 50%

Section B (p 39)

1 (a) £10   (b) £8.50   (c) £45

2 (a) £15   (b) £13   (c) £12.50

3 (a) 7 kg   (b) £7.50   (c) £7.30

4 (a) 5p   (b) 15p   (c) £4.85
(d) 90p   (e) £4.50

5 (a) Divide by 10 to find 10%.
(b) Divide by 10, then halve to get 5%.
(c) Add the results for 10% and 5% together.

6 (a) £1.50   (b) £6   (c) £13.50

7 (a) 3% of £10 is £0.30.
All the others are £3.

(b) 75% of £8 is £6.
All the others are £5.

(c) 5% of £80 is £4.
All the others are £4.50.

Section C (p 40)

1 (a) 0.68   (b) 0.41   (c) 0.8
  (d) 0.08   (e) 0.09   (f) 0.9
  (g) 0.545   (h) 0.305

2 (a) 65%   (b) 60%   (c) 6%
  (d) 6.5%   (e) 2%   (f) 22%
  (g) 20%   (h) 20.2%

3

$\frac{58}{100}$	=	**0.58**	=	**58%**
$\frac{32}{100}$	=	0.32	=	**32%**
$\frac{9}{100}$	=	**0.09**	=	**9%**
$\frac{80}{100}$	=	**0.8**	=	80%
$\frac{5}{100}$	=	0.05	=	**5%**

4 (a) 7% $\frac{8}{100}$ 0.09 0.1 0.8 $\frac{81}{100}$
  (b) 0.02 0.1 12% 20% $\frac{1}{2}$ 0.7
  (c) 0.07 $\frac{8}{100}$ 0.3 and $\frac{3}{10}$ 40% 75%

Section D (p 40)

1 28% of 4 = 1.12
47% of 220 = 103.4
91% of 7.3 = 6.643
8% of 8 = 0.64

The answer for the extra calculation is
26% of 82 = 21.32

2 A is the odd one out (£10).
The others are £6.

3 A is the odd one out (3.2 kg).
The others are 3.3 kg.

4 (a) £1199.25   (b) £6795.75

5 (a) £410.50 (b) £36 306 (c) £2856

Sections E and F (p 41)

1 $\frac{13}{27}$ (0.48), $\frac{6}{11}$ (0.55), $\frac{11}{19}$ (0.58)

2 76%

3 12%

4 (a) 31%   (b) 69%

5 B is the odd one out (24.7%). Others are 25%.

6 B is the odd one out (56%). Others are 60%.

7 (a) 8.7% (b) 88.4% (c) 2.9%

Sections G and H (p 42)

1 (a), (b) The percentages for the pie charts are

	Protein	Carbohydrate	Fat	Fibre
QO	13%	70%	9%	8%
AB	14%	50%	4%	32%

(c) The protein content of each are similar, but All Bran contains a lot more fibre and less carbohydrate and fat than Quaker Oats.

2 (a) The percentages for the pie charts are

	Coal	Petroleum	Natural gas	Nuclear
1982	34.7%	36.2%	23.0%	6.1%
1995	22.9%	35.1%	32.2%	9.8%

(b) The use of coal has dropped considerably. Natural gas and nuclear energy have increased their shares.

3 (a) 30p   (b) 13%

4 (a)

of	5	82	**50**
10%	0.5	**8.2**	5
20%	**1**	**16.4**	10
70%	3.5	**57.4**	35

(b)

of	30	55	**20**
30%	**9**	**16.5**	6
80%	**24**	**44**	16
20%	**6**	11	**4**

Think of a number

This unit develops the important idea of an inverse process and shows how to use it to solve an equation.

In later work, pupils should realise the limitations of using arrow diagrams to solve equations. The 'balancing' method should then become the principal method for solving equations.

In this unit, arrow diagrams are drawn with circles and ellipses. Pupils may find it easier to use squares and rectangles.

p 126 **A** Number puzzles	Using arrow diagrams and inverses to solve 'think of a number' puzzles
p 128 **B** Using letters	Linking equations, arrow diagrams and puzzles
p 129 **C** Solving equations	Solving linear equations by using arrow diagrams and inverses
p 130 **D** Quick solve game	A game to consolidate solving equations

---

Essential

Calculators
Sheet 157

Practice booklet pages 43 to 45

Optional

Sheet 158

A Number puzzles (p 126)

Calculators for working with decimals and large numbers

'This introduction went down very well.'
'It was a valuable run up to solving equations.'

◊ As a possible introduction, ask the pupils each to think of a number without telling anyone what it is. Now ask them to:

Add 1.

Multiply by 3.

Add 5.

Take away 2.

Divide by 2.

Ask some pupils to tell you what their answers are and then work backwards to give the numbers they were thinking of.
Pupils could discuss how they think you worked them out.

Now give some single-operation problems such as

'I think of a number.
I divide by 0.2 and my answer is 15.
What number did I think of?'

Ensure discussion of these brings out the idea of using an inverse operation in a reversed arrow diagram.

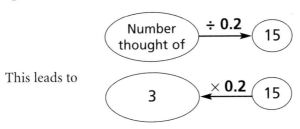

This leads to

Encourage pupils to check each solution by substituting in the original puzzle.

Now move on to 'think of a number' problems that use more than one operation and show how arrow diagrams and inverses can be used to solve them.

Some pupils will feel more confident using 'trial and improvement' to solve these problems. To demonstrate the power of using inverses, include problems involving decimals and many operations.

For example,

'I think of a number.
I divide by 5.
I subtract 45.
I multiply by 1.2.
I add 8.
I multiply by 0.2.
I subtract 12.
My answer is 10.
What number did I think of?'

Pupils could use both trial-and-improvement and inverse methods and compare them.

You could end this teacher-led session by asking the pupils to solve the puzzles on the pupil's page. The solutions to these puzzles are 18, 12 and 15.3 respectively.

A6 Pupils who are struggling with this can be reminded of work from 'Inputs and outputs'. Ask them to consider what happens if n is used as the 'input' for this puzzle.

'Some students were annoyed at going back to "Inputs and outputs", but I feel it was a good idea.'

B Using letters (p 128)

It may help to tell pupils that this section is not about solving the puzzles or equations, but about linking equations, arrow diagrams and puzzles. Otherwise, they may feel that they have not fully answered the questions unless they have solved each puzzle to find the number thought of.

The questions could be used as a basis for a whole-class discussion.

◊ Include some examples in your introduction where multiplication is the first operation and some where it is the second operation. Pupils must understand when they need to use brackets. You could include the puzzle and diagrams leading to the equation $2n + 3 = 20$, bringing out how it differs from $2(n + 3) = 20$ (the example on page 128).

◊ Make sure pupils know that $2 \times (n + 3)$ is the same as $(n + 3) \times 2$ and that $2(n + 3)$ is shorthand for it.

◊ Introduce the 'division line' and make sure pupils are aware that, for example,

$(n \div 4) + 5 = 7$ can be written $\dfrac{n}{4} + 5 = 7$

and $(n + 5) \div 4 = 7$ can be written $\dfrac{n+5}{4} = 7$.

C Solving equations (p 129)

Calculators for working with decimals and large numbers

'Pupils made up their own equations for others to solve.'

◊ Pupils who have already used balancing ideas may want to extend the method here. Point out that it is possible to use the balancing approach in this section but it needs to be modified to deal with equations involving subtraction. However, explain that using arrow diagrams to solve equations should improve their understanding of the general processes involved and help with later work.

◊ In your discussion, include the equation $2n - 3 = 130$ and compare it with $2(n - 3) = 130$ on the pupil's page.

◊ The equations in question C5 involve more than two operations. It may be beneficial to include some examples of this kind in your introduction.

D Quick solve game (p 130)

'This was worthwhile. It had a different feel – pupils were handling it in their heads.'

The version of this game described in the pupil material can be played in groups of three or four. It can also be played as a whole class or individually (see below).

> Each group needs a set of 36 cards (Sheet 157)
> Optional: Each pupil needs a copy of sheet 158 if they check each other's answers.

◊ Emphasise that **all** players take a card at the start of the game and take another as soon as they think they have solved their equation. They should keep their equation cards – they will not be used by another player.

◊ There are other ways to use the cards.

One pile version

This is played as the version in the book but pupils take cards from a mixed shuffled pile.

Three pile whole-class version

The game could be played as a whole class with the teacher having sets of cards in three piles: cards worth 1 point, 2 points and 3 points.

Individual pupils ask the teacher for a 1, 2 or 3 point card. When they think they have solved the equation, they ask for another card.

Continue until all the cards have been taken or some specified time limit has been reached.

One pile whole-class version

This is played as the Three pile whole-class version but pupils take cards at random from a mixed pile.

Individual version

The cards do not need to be cut out for this version. Each pupil solves as many equations as they can from the set of 36 in a specified time. Solutions are checked and points awarded as before.

The game can be played with the number of points awarded for a correct solution being the value of the solution.

A set of solutions is given below:

Card 1	$n = 3.5$	Card 13	$n = 5.2$	Card 25	$n = 4$
Card 2	$n = 1.5$	Card 14	$n = 4$	Card 26	$n = 0.6$
Card 3	$n = 2$	Card 15	$n = 29.5$	Card 27	$n = 19.3$
Card 4	$n = 9$	Card 16	$n = 0.6$	Card 28	$n = 5$
Card 5	$n = 18$	Card 17	$n = 5.8$	Card 29	$n = 0.7$
Card 6	$n = 0.6$	Card 18	$n = 2$	Card 30	$n = 1.7$
Card 7	$n = 2$	Card 19	$n = 6$	Card 31	$n = 4.48$
Card 8	$n = 4$	Card 20	$n = 8$	Card 32	$n = 35$
Card 9	$n = 5$	Card 21	$n = 2.7$	Card 33	$n = 3$
Card 10	$n = 3.3$	Card 22	$n = 6$	Card 34	$n = 143$
Card 11	$n = 108$	Card 23	$n = 0.9$	Card 35	$n = 21$
Card 12	$n = 241$	Card 24	$n = 1.5$	Card 36	$n = 34$

Ⓐ Number puzzles (p 126)

A1

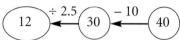

The number thought of was 33.

A2 (a) Puzzle 1 C; Puzzle 2 B

(b) Puzzle 1

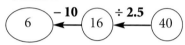

The number thought of was 12.

Puzzle 2

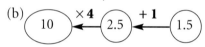

The number thought of was 6.

A3 (a) I think of a number.
 • I divide by 4.
 • I subtract 1.
 The result is 1.5.
 What was my number?

(b)

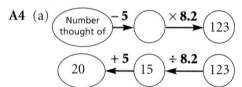

The number thought of was 10.

A4 (a)

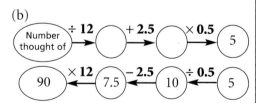

The number thought of was 20.

(b)

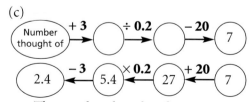

The number thought of was 90.

(c)

The number thought of was 2.4.

A5 Using positive whole numbers and zero gives the following possibilities.
 • I add 10, I multiply by 1.
 • I add 4, I multiply by 2.
 • I add 2, I multiply by 3.
 • I add 1, I multiply by 4.
 • I add 0, I multiply by 6.
Extending to negative numbers and decimals gives an infinite number of possibilities.

A6 (a) The pupil's numbers and solutions

(b) The pupil's numbers and solutions

(c) The number thought of is the same as the result each time.
It can be explained by writing 'n' for the 'number thought of' and writing a simplified expression in terms of n as the result of each operation.

Number thought of n

I subtract 1. $\quad n - 1$

I multiply by 6. $\quad 6(n - 1) = 6n - 6$

I add 3. $\quad 6n - 3$

I divide by 3. $\quad 2n - 1$

I add 1. $\quad 2n$

I divide by 2. $\quad n$

Ⓑ Using letters (p 128)

B1 (a) A Y, B W, C V, D X

(b)

$$n \xrightarrow{\times 5} \bigcirc \xrightarrow{+7} 16$$

B2 (a) W D, X B, Y E, Z C

(b) I think of a number.
 • I subtract 5.
 • I multiply by 4.
 The result is 8.
 What was my number?

B3 (a) $3n + 4 = 108$ (b) $4.5n = 162$

(c) $\dfrac{n-2}{5} = 2.2$

B4 (a) $\dfrac{n}{3} + 6 = 21$ (b) $2.6(k - 5) = 65$

ℂ **Solving equations** (p 129)

C1 (a) $n = 12$ (b) $h = 10$ (c) $p = 21$
   (d) $q = 71$ (e) $t = 16.8$ (f) $s = 19$

C2 (a) $n = 7.4$ (b) $p = 3.9$ (c) $q = 32$
   (d) $x = 3.6$ (e) $y = 22.1$ (f) $z = 9.6$

C3 $(6 - 3) \div 5 = 3 \div 5 = 0.6$; the pupil's
   equations with solution $n = 6$

C4 The pupil's equations with solution
   $y = 1.5$

C5 (a) $m = 23$ (b) $p = 0.7$
   (c) $s = 9.5$ (d) $t = 0.05$

What progress have you made? (p 130)

1 (a)
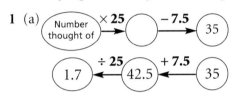

   The number thought of was 1.7.

(b)
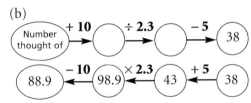

   The number thought of was 88.9.

2 A Z, B W, C Y

3 (a) $z = 20$ (b) $p = 18$ (c) $y = 96$
   (d) $q = 3.7$ (e) $x = 5.31$

Practice booklet

Section A (p 43)

1 (a) 32 (b) 9 (c) 2.8 (d) 248.4

2 (a) The pupil's result and solution
   (b) The pupil's solution

(c) The result is always twice the number
   thought of. It can be explained by
   writing 'n' for the 'number thought
   of' and writing a simplified
   expression in terms of n as the result
   of each operation.

Number thought of n
I add 1 $n + 1$
I multiply by 4 $4(n + 1) = 4n + 4$
I add 6 $4n + 10$
I divide by 2 $2n + 5$
I subtract 5 $2n$

Section B (p 43)

1 $54n = 378$

2 (a) $7n + 3 = 206$ (b) $\dfrac{n - 13}{4} = 6.1$

3 (a) $7(y - 3) = 21$ (b) $\dfrac{a}{5} + 3 = 18$
   (c) $\dfrac{3.2p}{7} = 4$ (d) $5.1(m + 9) = 102$
   (e) $\dfrac{x - 10}{3} + 1 = 3$

Section C (p 44)

1 (a) $w = 11$ (b) $n = 8$ (c) $p = 31$
   (d) $x = 15.3$ (e) $a = 77$ (f) $y = 39$
   (g) $p = 23$ (h) $h = 16.2$ (i) $t = 152$

2 (a) $x = 5.1$ (b) $a = 4.3$ (c) $y = 50$
   (d) $m = 4.2$ (e) $w = 15.3$ (f) $h = 11.5$
   (g) $k = 13.5$ (h) $m = 216.9$
   (i) $n = 46.4$

3 The pupil's equations with solution $x = 4$

4 The pupil's equations
   with solution $p = 2.5$

5 The pupil's equations
   with solution $n = 7.1$

6 (a) $x = 16$ (b) $d = 14$ (c) $y = 17.9$
   (d) $m = 7$ (e) $b = 16.7$ (f) $n = 16$

*7 (a) $s = 12.9$ (b) $r = 1.5$
   (c) $q = 2.4$ (d) $p = 8.5$

*8 $x = 8$

Quadrilaterals

The names and properties of the square, rectangle, parallelogram, rhombus, trapezium, kite and arrowhead are revised or introduced. Pupils see how special types of quadrilateral can be made up from certain types of triangle. The sum of the interior angles of a quadrilateral is established and used to find missing angles and as part of an exercise on accurate drawing. The fact that some types of quadrilateral are special cases of others is briefly explored.

Essential

Square dotty paper, sheet 164
Scissors, angle measurers, compasses

Practice booklet pages 46 to 47

A Special quadrilaterals (p 131)

> Square dotty paper

◊ Page 131 provides an opportunity to lead a discussion to find out how many of the quadrilaterals' names and properties pupils know and to fill in any gaps in their knowledge.

One teacher reported 'I organised pupils into groups of four or five and gave them 30 minutes to prepare a presentation for the class on their particular shape. Each contained an accurate drawing of the shape and responses to the seven questions on the page. This worked well!'

A6–8 These questions bring in area and are appreciably harder than the rest.

Ⓑ Quadrilaterals from triangles (p 134)

> Sheet 164, scissors

Although this section takes time it provides consolidation: pupils have to recognise triangle types and the special quadrilaterals they produce. It also offers practice in exploring all possibilities and can help pupils visualise quadrilaterals as built up from triangles. Most parts of B8 (and of section E) depend on this last idea.

◊ You may need to revise the different types of triangles before starting this section. Remember that some types of quadrilaterals are special cases of others (a point dealt with more fully later). So, for example, if a pupil creates a rhombus in B3 and labels it 'parallelogram' that is not wrong, but you could ask 'What special kind of parallelogram?'

Ⓒ Angles of a quadrilateral (p 135)

> Angle measurers

◊ You could remind pupils of the proof that the angles of a triangle add up to 180°, in either of these versions

By splitting up a quadrilateral into two triangles, we *prove* that its angles add up to 360°. Make sure that pupils appreciate the difference between proving and merely verifying by measurement.

Ⓓ Accurate drawing (p 138)

> Angle measurers, compasses

Ⓔ Quadrilaterals from diagonals (p 138)

◊ Some pupils may need reassurance that the sketch is not trying to indicate the *shape* of any of the quadrilaterals: it merely shows how the vertices are lettered.

F Stand up if your drawing … (p 139)

◊ It's probably best to start this section with the pupils' books shut. We define a parallelogram as any quadrilateral that has two pairs of parallel sides. So a rectangle is a special kind of parallelogram.

Similarly,
a parallelogram is a special kind of a trapezium,
a rhombus is a special kind of a parallelogram,
a square is both a special rhombus and a special rectangle.

Hence,
a square is a special parallelogram,
a rhombus is a special trapezium,
and so on.

Introduce these ideas through discussion before going on to the following activity.

Six pupils sit on chairs facing the front of the class each holding one of these drawings.

(isosceles trapezium, trapezium with two right angles, and non-special cases of these: rectangle, parallelogram, rhombus, square).

Say to the group 'Stand up if your drawing is a rectangle.' If the square person doesn't stand up, ask the class if there is anyone not standing up who should be and, by inviting explanations, check that the special case idea has been understood.

Repeat the process with 'Stand up if your drawing …
        … is a rhombus'
        … is a square'
and so on. You can also extend the idea to properties of their drawings:
'Stand up if your drawing …
        … has reflection symmetry'
        … has at least one right angle'
and so on.

After you have done these activities, pupils can decide what the teacher asked for in the photographs shown on the page.

G Always, sometimes, never … (p 139)

◊ Each numbered box describes a type of quadrilateral. Through teacher-led discussion, the class has to decide which. The discussion should not be hurried: pupils should have time to suggest quadrilaterals and comment on others' suggestions.

Ⓐ Special quadrilaterals (p 131)

A1

A2 (a)

Square

(b)

Rhombus

(c)

Rhombus

(d)

Parallelogram

(e)

Rhombus

(f)

Parallelogram

A3

(a)   (b)

A4

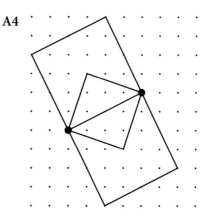

A5 The pupil's drawing of a rhombus
(There are infinitely many possible.)

A6

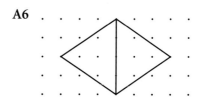

A7 Here are two possibilities. There are others.

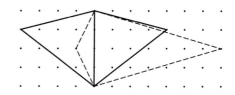

A8

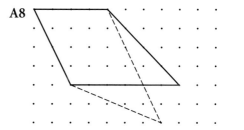

B Quadrilaterals from triangles (p 134)

B1 These quadrilaterals are possible.

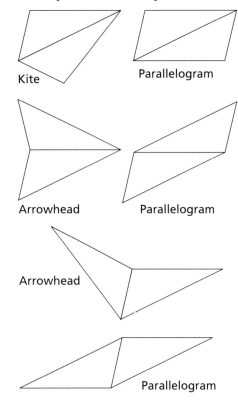

Kite

Parallelogram

Arrowhead

Parallelogram

Arrowhead

Parallelogram

B2 You could not make an arrowhead.

B3 These quadrilaterals are possible.

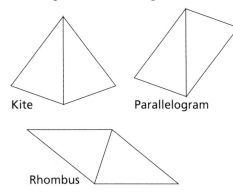

Kite

Parallelogram

Rhombus

B4 A kite would not have been possible but an arrowhead would.

B5 Only this rhombus is possible.

B6 (a) These quadrilaterals are possible.

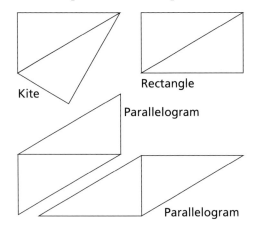

Kite

Rectangle

Parallelogram

Parallelogram

(b) These triangles are possible.

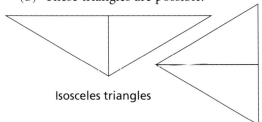

Isosceles triangles

B7 The rectangle and kite would have become a square. Only one isosceles triangle would have been possible, a right-angled one.

B8 (a) $\frac{1}{3}$ (b) $\frac{1}{4}$ (c) $\frac{1}{2}$
    (d) $\frac{3}{8}$ (e) $\frac{1}{7}$

C Angles of a quadrilateral (p 135)

C1 360°

C2 $a = 140°$ $b = 72°$ $c = 65°$

C3 $a = 222°$ $b = 100°$ $c = 112°$
    $d = 115°$ $e = 65°$ $f = 30°$
    $g = 92°$ $h = 140°$ $i = 105°$
    $j = 67°$ $k = 95°$ $l = 80°$
    $m = 55°$ $n = 100°$

C4 $a = 87°$ $b = 101°$ $c = 152°$ $d = 84°$

C5 $a = 135°$ $b = 45°$ $c = 60°$ $d = 45°$
    $e = 30°$ $f = 30°$ $g = 20°$ $h = 15°$
    $i = 30°$ $j = 24°$ $k = 18°$

D Accurate drawing (p 138)

D1 The pupil's drawings

D2 (a) 5.1 cm   (b) 14.6 cm   (c) 11.1 cm

D3 (a) 58°   (b) 206°

D4 The pupil's drawing

D5 The pupil's drawing

D6 The pupil's drawing

E Quadrilaterals from diagonals (p 138)

E1 (a) A square   (b) A kite
(c) A rectangle   (d) A rhombus
(e) A parallelogram
(f) An ('isosceles') trapezium
(g) A (long thin) rectangle
(h) A parallelogram

F Stand up if your drawing ... (p 139)

The teacher said:

1 'Stand up if your drawing is a parallelogram.'
2 'Stand up if your drawing has just two lines of reflection symmetry.'
3 'Stand up if your drawing is a trapezium' (or, possibly, '... if your shape is a quadrilateral').

G Always, sometimes, never ... (p 139)

1 A rhombus
2 A rectangle
3 A square
4 A kite
5 A rhombus
6 A trapezium
7 A kite
8 A parallelogram

What progress have you made? (p 140)

1 The pupil's drawings

2 A kite, an arrowhead

3 A square

4 $a = 92°$ $b = 240°$ $c = 55°$

5 The pupil's drawings

6 Square, rhombus, rectangle

Practice booklet

Section B (p 46)

1 The pupil's drawings, for example

(a)   (b)

(c) 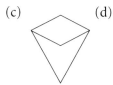   (d)

(e)

Section C (p 46)

1 $a = 98°$ $b = 117°$ $c = 102°$
$d = 48°$ $e = 95°$ $f = 68°$
$g = 244°$ $h = 16°$ $i = 144°$
$j = 81°$

2 $a = 36°$ $b = 20°$ $c = 28°$
$d = 18°$ $e = 25°$ $f = 20°$

Section D (p 47)

1 The pupil's accurate drawings

⑱ Negative numbers

This unit revises addition and subtraction of negative numbers and introduces multiplication and division.

Practice booklet page 48

A Addition and subtraction patterns (p 141)

This approach to addition and subtraction complements the approach used in *Book 1*. The idea of extending number patterns is used for multiplication in the next section.

◊ You could start by asking pupils to imagine that they do not know any rules for adding or subtracting negative numbers (which may well be the case anyway!).

Complete the next two lines of pattern A and notice that the result goes down by 1 each time. Use this fact to extend the pattern and add two or three more lines beyond those given.

Now ask pupils, working in pairs, to do the same for each of the other patterns. Help them to appreciate that the rules for adding and subtracting negative numbers are chosen so that the results 'fit in' with those for positive numbers.

B Multiplication (p 142)

◊ It is probably best to reproduce the incomplete table on the board or an OHP transparency.

You could start by asking pupils how the table should be extended to the left. In row '4', for example, the numbers go down by 4 each time as you go to the left, so the row continues $^-4, ^-8, \ldots$

Then work on extending the columns downwards, starting with columns '4', '3', …

◊ Point out the square numbers on the diagonal starting with 16. Each square number (apart from 0) appears twice in the table and thus has two square roots, one positive and one negative.

◊ You could also ask high attainers to think about the consequences of choosing a different rule for multiplying two negative numbers. Most pupils will see that $3 \times ^-2$ can be thought of as $^-2 + ^-2 + ^-2 = ^-6$. However, if $^-3 \times ^-2$ were also $^-6$, then $^-6 \div ^-2$ would have two answers, 3 or $^-3$.

ⓒ **Division** (p 143)

ⓓ **Using inverse operations** (p 144)

ⒶAddition and subtraction patterns (p 141)

A1 (a) 5 (b) 9 (c) 16
(d) ⁻4 (e) 16

A2 (a) 4 (b) ⁻5 (c) 5
(d) ⁻5 (e) 3

A3 (a) ⁻4 (b) 1 (c) ⁻16
(d) 13 (e) 3

A4 Each of the additions below can be written as another addition and in two ways as a subtraction.

$$^-2 + 3 = 1 \quad ^-4 + 7 = 3$$
$$^-5 + 1 = ^-4 \quad ^-5 + 3 = ^-2$$

There are 16 possibilities altogether.

Ⓑ Multiplication (p 142)

B1 (a) ⁻21 (b) 21 (c) 30
(d) ⁻24 (e) 24

B2 (a) ⁻30 (b) 24 (c) ⁻30 (d) ⁻48

B3 ⁻3

B4 32

B5 (a) ⁻15 (b) ⁻12 (c) 23
(d) ⁻36 (e) 19 (f) 34

B6 Clare was not right. As well as 8, Kirsty's number could have been ⁻8.

B7 4 or ⁻4

*B8 6 or ⁻8

ⒸDivision (p 143)

C1 ⁻20 ÷ 4 = ⁻5, ⁻20 ÷ ⁻5 = 4

C2 18 ÷ ⁻6 = ⁻3, 18 ÷ ⁻3 = ⁻6

C3 negative/positive = negative
positive/negative = negative
negative/negative = positive

C4 (a) ⁻4 (b) 10 (c) ⁻4 (d) ⁻5
(e) 8 (f) 4 (g) ⁻7 (h) ⁻5

C5 14

C6 (a) ⁻9 (b) ⁻11 (c) 16 (d) 21
(e) 7 (f) 2

C7 (a) ⁻2

(b) (i) + 6, ÷ ⁻3, + 3, × ⁻2
(ii) ÷ ⁻3, × ⁻2, + 3, + 6
(or + 6, + 3)
(iii) × ⁻2, + 6, + 3, ÷ ⁻3
(or × ⁻2, + 3, + 6, ÷ ⁻3)
(iv) + 3, × ⁻2, + 6, ÷ ⁻3
(or ÷ ⁻3, + 3, × ⁻2, + 6)

C8 (a) $\dfrac{24}{3} - \dfrac{^-12}{^-4}$ (b) $\dfrac{^-12}{3} - \dfrac{24}{^-4}$

(c) $\dfrac{^-12}{^-4} - \dfrac{24}{3}$ (d) $\dfrac{24}{^-12} - \dfrac{3}{^-4}$

Ⓓ Using inverse operations (p 144)

D1 (a) 25
(b) (i) ⁻7 (ii) 2 (iii) 6

D2 (a) ⁻4
(b) (i) ⁻31 (ii) 17 (iii) 85

D3 (a) ⁻7 (b) 2 (c) ⁻8

D4 (a) ⁻4 (b) 25 (c) ⁻14

What progress have you made? (p 144)

1 (a) 12   (b) ⁻9   (c) ⁻6

2 (a) 28   (b) ⁻27   (c) ⁻16

  (d) ⁻3   (e) ⁻8   (f) ⁻5

3 (a) ⁻2   (b) 5

Practice booklet

Sections A and B (p 48)

1 (a) ⁻13   (b) ⁻6

  (c) 9   (d) 5

2 (a) ⁻5   (b) 15   (c) 13

  (d) ⁻15   (e) 9

3 (a) ⁻24   (b) ⁻25   (c) 27

  (d) ⁻19   (e) 44

4 3 or ⁻3

5 5 or ⁻5

Sections C and D (p 48)

1 (a) ⁻1   (b) 7   (c) ⁻4

  (d) 14   (e) ⁻5

2 (a) ⁻6   (b) 2   (c) 17

  (d) 14   (e) ⁻1

3 (a) ⁻1   (b) ⁻5   (c) 3

4 (a) 6   (b) 1

Fair to all?

The mean is introduced using the idea of fairness.
There is some practical work in the unit, and a section that brings
together mean, mode and median.

p 145	**A** How to be fair	Calculating means
p 148	**B** Mean from frequencies	Calculating a mean from a frequency table
p 151	**C** Words	An investigation
p 152	**D** Averages	Calculating all three averages: mean, median and mode
p 153	**E** Mean challenges	Solving problems about means

Essential	**Optional**
Packs of playing cards (ace to 10 only)	Newspapers Foreign language texts, other texts
Practice booklet pages 49 to 52	

A How to be fair (p 145)

◊ You could start by asking pupils in pairs to decide whether Sharon's or Joshua's group did better, and why.

The mean for Sharon's group is 8 kg and for Joshua's it is 7 kg.

A6 This can generate discussion! Using the median reverses the order, and pupils may bring in the idea of consistency.

Emphasise that the mean is not necessarily a possible number of points.

Mean tricks (p 147)

Packs of playing cards (ace to 10 only) or equivalent

◊ This game not only consolidates mental skills in calculating the mean but can bring out other ideas (for example that the deviations from the mean add up to zero). Since any number of cards (up to 7) can be used when working out the mean, each hand potentially contains many calculations of a mean.

Eavesdrop groups of pupils to hear their methods of finding the mean. There could be class discussion of these methods. Schools have found that the benefits pupils get from this game improve with playing as they develop strategies.

B Mean from frequencies (p 148)

◊ It is worth spending some time on the meaning of the word 'frequency' in each situation. It is a common misconception to find the mean of the frequencies.

◊ Encourage pupils to set out their work logically. Although there are subroutines on a calculator, at this stage it is better to show all the steps so that errors can be traced.

B12 A common error is to assume that the mean for the whole population is equal to the mean of the means for the two parts.

Investigation

| Newspapers (*Mirror* and *Guardian* suggested), foreign language texts |

◊ Samples need to be of a reasonable size, say 20 to 30 sentences in each paper. In some schools this worked well as a homework task. Pupils were allowed to choose from a wider range of newspapers however.

C Words (p 151)

| Optional: other texts |

◊ This page is deliberately obscured so that pupils cannot count every word! Some schools have used this successfully on books of their own choice. The activity takes about 15–20 minutes. If the piece is word-processed the teacher could use a word count.

D Averages (p 152)

◊ This section brings together the three measures of average met so far. The word 'average' is used loosely to mean 'about the middle'. For 'Mary is of average height' you could ask:

• Is 'average height' the average for the school year? the school? the population of the UK?

• What average could it be?

The 'average maximum daily temperature' would usually be understood as the mean of the maximum daily temperatures (probably rounded to the nearest degree).

In the sentence about the 'average family', the word 'average' has been attached to 'family' but really refers to an average amount of water.

E Mean challenges (p 153)

E1 It is worth discussing different ways of calculating an answer. For example, you could find the total weight before and after replacement or you could divide the increase in weight by the number in the team.

A How to be fair (p 145)

A1 $27 \div 9 = 3$

A2 $120 \div 10 = 12$ goldfish

A3 (a) $40 \div 5 = 8$ pupils
(b) $36 \div 4 = 9$ pupils

A4 7 peppers per plant

A5 Ruth is right. The 24 peppers have to be shared between the four plants.

A6 Pat $32 \div 4$ or 8 points per game
Jon $39 \div 5$ or 7.8 points per game
Wayne $51 \div 6$ or 8.5 points per game

So you might consider Wayne to be the best points scorer.

A7 (a) 8L $40 \div 20 = 2$ cans
8N $55 \div 25 = 2.2$ cans
8N gets the prize.
(b) 8L $50 \div 20 = 2.5$ bottles
8N $60 \div 25 = 2.4$ bottles
8L gets the prize.
The pupil's explanation
(c) The pupil's own view and explanation
(8L average items = 4.5)
(8N average items = 4.6)

A8 Probably the best way to answer these questions is to calculate the mean load pulled by a member of the team.

Team A
mean load pulled = $2475 \div 9 = 275$ kg

Team B
mean load pulled = $1656 \div 6 = 276$ kg

Team C
mean load pulled = $1911 \div 7 = 273$ kg

(a) Team B   (b) Team C
The pupil's reasons

B Mean from frequencies (p 148)

B1 (a) 5   (b) 15
(c) 1156 kg   (d) 77.1 kg

B2 (a) 34   (b) 124   (c) 3.6

B3 $(3 \times 4) + (4 \times 8) + (5 \times 5) + (6 \times 4) + (7 \times 3) + (8 \times 7) + (9 \times 5) = 215$
$215 \div 36 = 6.0$ (to 1 d.p.)

B4 (a) $(0 \times 5) + (1 \times 12) + (2 \times 13) + (3 \times 8) + (4 \times 2) = 70$
(b) $70 \div 40 = 1.75$

B5 $233 \div 40 = 5.825$

B6 $(8 \times 2) + (7 \times 3) + (6 \times 9) + (5 \times 1) = 96$
$96 \div 15 = 6.4$

B7 $222 \div 104 = 2.1$ (to 1 d.p.)

B8 $1180p \div 20 = 59p$

B9 $1435 \div 50 = 28.7$

B10 $8938 \div 177 = 50.5$ (to 1 d.p.)
The statement seems fair. Although you might only get 48 matches, it averages out at more than the 50 stated.

B11 (a) The mean of the pupil's set of five numbers
(b) The mean increases by 2.
(c) The mean increases by the amount you add to each number.
(d) $9760 + [(1 + 3 + 0 + 2 + 4) \div 5] = 9762$

B12 (a) $265 \div 20 = 13.25$

(b) $420 \div 30 = 14$

(c) $685 \div 50 = 13.7$

Ⓒ Words (p 151)

C1 One way is to find the mean number of words in a line, then multiply this by the number of lines.
Number of words in first six lines
18, 17, 17, 17, 18, 16

Mean number of words in first six lines
$= 103 \div 6 = 17. \ldots$

Maximum number of words (assuming all lines of full length)
$= 17.16\ldots \times 32 = 549.3$ (to 1 d.p.)

Minimum number of words (assuming some lines are very short)
$= 17.16\ldots \times 28 = 480.7$ (to 1 d.p.)

So anything between 550 and 481 is reasonable, e.g. 515.

C2 $5000 \div 515 = 9.7$ (approx.)
i.e. about 9 or 10 pages

Ⓓ Averages (p 152)

D1 (a) Median   (b) Mode

D2 (a) Median 23, mean 26.5 (to 1 d.p.), mode 18

(b) The frequencies are too low for the modal age to be relevant.

(c)

Group	10–19	20–29	30–39	40–49	50–59
Frequency	11	9	6	3	1

The modal age group is 10–19.

D3 *Blackmouth*
Median   5
Mean $189 \div 31 = 6.1$ (to 1 d.p.)
Mode   4 and 5 equal

Bournepool
Median   7
Mean $178 \div 31 = 5.7$ (to 1 d.p.)
Mode   7

Although the mean number of hours of sunshine is less for Bournepool (5.7 compared with 6.1), the median is higher. The mode of 7 hours at Bournepool is more than that of the two most frequent amounts of sunshine at Blackmouth (4 and 5 hours).

Ⓔ Mean challenges (p 153)

E1 Total weight before $= 7 \times 58 = 406$
after $= 406 - 40 + 54 = 420$
new mean $= 420 \div 7 = 60\,\text{kg}$
or
Increase in weight $= 14\,\text{kg}$
so increase in mean $= 14 \div 7 = 2\,\text{kg}$
new mean $= 58 + 2 = 60\,\text{kg}$

E2 Increase $= 7 \times 2 = 14\,\text{kg}$
so new player's weight $= 43 + 14 = 57\,\text{kg}$

E3 $(4 \times 143 + 140) \div 5 = 676 \div 5 = 135.2\,\text{cm}$

E4 $(6 \times 45) + (5 \times 51) = 525$
$525 \div 11 = 47.7\,\text{kg}$ (to 1 d.p.)
Note that it is not the mean of the two means.

E5 $(5 \times 46) - (40 \times 4) = 70$

What progress have you made? (p 153)

1 $94 \div 7 = 13.4°\text{C}$ (to 1 d.p.)

2 $70 \div 25 = 2.8$ eggs

Practice booklet

Section A (p 49)

1 (a) Kathy did better.
Her mean score was 3.5,
Mark's mean score was 3.4.

(b) Mark would have done better.
His mean would then be $16 \div 4 = 4$
skittles.

2 (a) The girls' mean pocket money is
£40.00 ÷ 10 = £4.00.

(b) The boys' mean pocket money is
£31.60 ÷ 8 = £3.95.

3 $52 \div 13 = 4$ letters per word

4 48.5p

5 (a) The pupil's answer, for example
5, 7, 9 or 3, 5, 13

(b) The total of the three numbers must
be 21 which is an odd number.
When even numbers are added the
answer is always even.

6 The pupil's answer, but the total of the
five numbers must be 47

Section B (p 50)

1 $(3 \times 28) + (1 \times 26) = 110$
$110 \div 4 = 27.5$ teeth

2 $(1 \times 50) + (2 \times 25) + (3 \times 20) + (4 \times 3)$
$+ (5 \times 2) = 182$
$182 \div 100 = 1.82$ people per car

3 $(4 \times 5) + (5 \times 1) + \ldots + (14 \times 4) = 440$
$460 \div 50 = 9.2$ eggs per lay

4 (a) $5370 \div 25 = 214.8$ grams

(b)
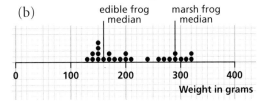

(c) 10

(d) $2880 \div 10 = 288$ grams

(e) $2490 \div 15 = 166$ grams

Section E (p 51)

1 $(10 \times 47.2) + (20 \times 42.4) = 1320$
$1320 \div 30 = 44\,\text{kg}$

2 (a) $15.43 \div 8 = 1.93$ m

(b) The pupil's estimate

(c) $16.89 \div 9 = 1.88$ m

3 (a) $66 \div 10 = 6.6$ goals

(b) $61 \div 10 = 6.1$ goals

(c) No
The answers to parts (a) and (b)
would suggest otherwise.

(d) Mean goal margin = 3.7 goals
The answers to parts (a) and (b) do
not predict this.

4 The one who joined was 18 years
younger than the one who left.
$(42 \times 6 - 39 \times 6 = 18)$

*5 There are eight sets:
1, 2, 3, 5, 6, 11, 14
1, 2, 3, 5, 7, 10, 14
1, 2, 3, 5, 8, 9, 14
1, 2, 4, 5, 6, 10, 14
1, 2, 4, 5, 7, 9, 14
1, 3, 4, 5, 6, 9, 14
1, 3, 4, 5, 7, 8, 14
2, 3, 4, 5, 6, 7, 15

Review 2 (p 154)

1 (a) $5a = 20 = 5(a + 4)$

(b) $6b - 12 = 6(b - 2)$

(c) $6c - 20 = 2(3c - 10)$

(d) $15d + 24 = 3(5d + 8)$

2 (a) 400   (b) 0.036

(c) 8.1   (d) $30 \times 0.04 = 1.2$

3 $a = 66°, \; b = 66°, \; c = 73°, \; d = 107°$

(all with reasons)

4 Alex £580.50, Bharat £464.40, Chris £245.10

5 (a) 13   (b) 1.4

6 Here is one possibility for each.

(a)   (b)

(c)

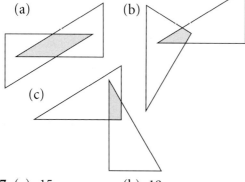

7 (a) 15   (b) 19

8 (a) Red 30%, blue 24%, green 11%, white 22%, black 8%, other 5%

(b) The pupil's pie chart

9 (a) C, D (b) A, F (c) none

10 (a) $3x - 1$ (b) $15y^2$ (c) $p - 8$

11 3.1

12 (a) ⁻4 (b) 29 (c) ⁻1

(d) 9 (e) ⁻5

13 $a = 102°, \; b = 103°, \; c = 116°, \; d = 112°$

(all with reasons)

14 551 is not prime. $551 = 19 \times 29$

15 (a) 53° (b) 75° (c) 49°

16

Fraction	$\frac{4}{5}$	$\frac{7}{100}$	$\frac{12}{25}$	$\frac{13}{20}$	$\frac{17}{50}$	$\frac{1}{20}$	$\frac{1}{8}$
Decimal	0.8	0.07	0.48	0.65	0.34	0.05	0.125
Percentage	80%	7%	48%	65%	34%	5%	12.5%

17 420

18 (a) $125 \, \text{cm}^2$ (b) $271 \, \text{cm}^2$

19 The pupil's drawings

Angle B = 94°, angle C = 101°, angle D = 57°, angle R = 65°, PS = 6.5 cm, SR = 5.8 cm

Mixed questions 2 (practice booklet p 53)

1 (a) $3p + 15$ (b) $4(q - 3)$

(c) $5r - 2$ (d) $\frac{1}{2}s + 7$

2 (a) 245 (b) 2.597 (c) 0.025 97

(d) 10.6 (e) 0.245 (f) 0.106

3 138°

4 (a) £3.25 (b) £9.60 (c) £10.50

(d) £8.40 (e) £17.10 (f) £5.00

5 (a) Trapezium, kite

(b) Rectangle, rhombus, square

6 (a) ⁻9   (b) 12

7 79.4 kg

8 42 m

9 (a) $x = 24$ (b) $x = 5.5$ (c) $x = 128$

(d) $x = 5$ (e) $x = ⁻3$ (f) $x = ⁻8$

10 15

② Know your calculator

The priority rules for multiplication, division, addition and subtraction are introduced by investigating how a scientific calculator evaluates expressions. The unit also covers the use of brackets, memory, square and square root and the 'change sign' key.

Scientific calculators are essential.

Essential

Scientific calculators (one for each pupil, for all sections of the unit)

Practice booklet pages 55 to 58

Ⓐ Order of operations (p 157)

◊ You could begin by asking pupils to predict what they think their calculators will give for each set of key presses on page 157.

Pupils find the result of each set of key presses (remind them they need to press the '=' key or 'ENTER' key at the end) and try to describe the rules they think the calculator uses to evaluate expressions that include any combination of the four operations. They should try their rules out on their own expressions. Working in groups, each group could try to produce a clear statement of the rules they think the calculator follows.

A brief statement of the rules could be:

• You multiply or divide before you add or subtract.

• Otherwise, work from left to right.

◊ Some calculators use the symbols * and / for × and ÷ respectively.

A1–2 A calculator can be used to check results.

A3 Pupils could consider how many different results are possible.
They could make up their own puzzles like this for someone else to solve.

Ⓑ Brackets (p 158)

Ⓒ A thin dividing line (p 158)

D Complex calculations (p 160)

The memory or an intermediate '=' can be used instead of brackets. The choice is a purely personal one.

E Squares (p 160)

In more complex calculations such as those at the end of E7, it is not necessary for pupils to use a continuous string of key presses (although some might like the challenge!). Some pupils will feel much more confident if they can write down intermediate steps.

F Square roots (p 162)

G Negative numbers (p 162)

A Order of operations (p 157)

A1 (a) 24   (b) 33   (c) 43   (d) 6
     (e) 5   (f) 9   (g) 16   (h) 25

A2 (a) 5   (b) 6   (c) 20   (d) 10
     (e) 10   (f) 4   (g) 32   (h) 12
     (i) 5

A3 (a) $24 \times 8 - 2$ (b) $24 + 8 \div 2$
     (c) $24 - 8 \times 2$ (d) $24 \div 8 \div 2$

B Brackets (p 158)

B1 (a) 35   (b) 47   (c) 5   (d) 11
     (e) 5   (f) 3   (g) 5   (h) 7

B2 (a) 11   (b) 3   (c) 6   (d) 3
     (e) 11   (f) 9   (g) 2   (h) ⁻1
     (i) 27

B3 (a) No   (b) Yes   (c) No   (d) Yes
     (e) No   (f) Yes   (g) No   (h) Yes
     (i) No   (j) No   (k) Yes   (l) No

C A thin dividing line (p 158)

C1 A and C
     B and G
     D and F
     H and I

C2 (a) $60 + \dfrac{3}{2}$ (b) $\dfrac{23 - 7}{4}$

     (c) $\dfrac{100 + 5}{3}$ (d) $4 - \dfrac{6}{5}$

     (e) $\dfrac{5}{3 - 1}$ (f) $\dfrac{18}{9} - 1$

     (g) $4 + \dfrac{12}{5 - 2}$

C3 (a) 7   (b) 12.5 or $12\frac{1}{2}$
     (c) 1   (d) 4
     (e) 5   (f) 2
     (g) 7   (h) 10

C4 (a) 5   (b) 5   (c) 7
     (d) 35   (e) 5

C5 (a) 4   (b) 7

C6 (a) 1.5   (b) 60   (c) 5

C7 (a) (i) 19   (ii) 27

       (iii) 9   (iv) 5

    (b) The pupil's description

C8 (a) 21   (b) 5   (c) 2

C9 A and C

C10 $\boxed{7}\ \boxed{-}\ \boxed{1}\ \boxed{=}\ \boxed{\div}$
$\boxed{(}\ \boxed{2}\ \boxed{+}\ \boxed{1}\ \boxed{)}$

or

$\boxed{(}\ \boxed{7}\ \boxed{-}\ \boxed{1}\ \boxed{)}\ \boxed{\div}$
$\boxed{(}\ \boxed{2}\ \boxed{+}\ \boxed{1}\ \boxed{)}$

Ⓓ **Complex calculations** (p 160)

D1 (a) 19.629   (b) 2.125   (c) 17

    (d) 10.4

D2 (a) 11.25   (b) 9   (c) 2.5

    (d) 22.1   (e) 3.8   (f) 0.5

    (g) 3.9   (h) 3   (i) 31

Ⓔ **Squares** (p 160)

E1 (a) 25   (b) 50   (c) 42

    (d) 25   (e) 145   (f) 12

    (g) 2   (h) 4

E2 (a) 100   (b) 400   (c) 2

    (d) 9   (e) 8

E3 (a) $\boxed{5}\ \boxed{\times}\ \boxed{7}\ \boxed{x^2}$

    (b) $\boxed{(}\ \boxed{5}\ \boxed{+}\ \boxed{7}\ \boxed{)}\ \boxed{x^2}$

    (c) $\boxed{1}\ \boxed{0}\ \boxed{0}\ \boxed{\div}\ \boxed{5}\ \boxed{x^2}$

E4 (a) 72   (b) 1   (c) 81   (d) 2

    (e) 14   (f) 33   (g) 6.5

E5 B, D and E

E6 $\boxed{1}\ \boxed{5}\ \boxed{+}\ \boxed{9}\ \boxed{x^2}\ \boxed{=}$
$\boxed{\div}\ \boxed{4}\ \boxed{x^2}$

or

$\boxed{(}\ \boxed{1}\ \boxed{5}\ \boxed{+}\ \boxed{9}\ \boxed{x^2}\ \boxed{)}$
$\boxed{\div}\ \boxed{4}\ \boxed{x^2}$

E7 (a) 5.76   (b) 5   (c) 1.5

    (d) 36   (e) 90   (f) 7.44

    (g) 3   (h) 10.5   (i) 10.388

E8 $\boxed{5}\ \boxed{\times}\ \boxed{6}\ \boxed{2}\ \boxed{+}\ \boxed{4}\ \boxed{\div}\ \boxed{3}$

Ⓕ **Square roots** (p 162)

F1 (a) 5   (b) 19   (c) 48   (d) 2

F2 (a) 3.5   (b) 11.56

    (c) 6.002   (d) 7

    (e) 18.7   (f) 27.5625

F3 One possibility is given here for each part.

    (a) $9 + \sqrt{4} \div 1$ (b) $9 + 4 \div \sqrt{1}$

    (c) $1 + 9 \div \sqrt{4}$ (d) $\sqrt{9} + 1 \div 4$

    (e) $1 + \sqrt{9} \div 4$ (f) $9 + 1 \div \sqrt{4}$

Ⓖ **Negative numbers** (p 162)

G1 The pupil's key presses

G2 (a) ⁻3.75   (b) ⁻11.25

    (c) 0.8375   (d) ⁻8.1

G3 (a) 64.1   (b) 59.6

G4

What progress have you made? (p 163)

1 (a) 39 (b) 38 (c) 2 (d) 1

2 (a) 33.4 (b) 18
  (c) 0.523 437 5 (d) 5

3 (a) 27.76 (b) 4.5

4 (a) 24.04 (b) ⁻5

Practice booklet

Section A (p 55)

1 (a) 12 (b) 50 (c) 27 (d) 2
  (e) 12 (f) 18 (g) 12 (h) 11
  (i) 29

2 (a) 4 (b) 3 (c) 6
  (d) 6 (e) 12 (f) 5
  (g) 7 (h) 33 (i) 3

3 (a) $36 \div 9 \times 3$ (b) $36 - 9 \div 3$
  (c) $36 - 9 \times 3$ (d) $36 \times 9 \div 3$

4 (a) $16 + 8 \div 4$ (b) $16 + 4 \times 8$
   or $8 \div 4 + 16$ or $4 \times 8 + 16$
  (c) $8 + 16 \div 4$ (d) $4 \times 8 \div 16$
   or $16 \div 4 + 8$ or $8 \div 16 \times 4$
   or $16 - 8 + 4$ or $4 - 16 \div 8$

Sections B and C (p 56)

1 (a) $3 \times 3 \times 4$ (b) $4 \times (5 + 4)$
  (c) $4 + 4 \times 8$ (d) $(20 - 2) \times 2$
  (e) $24 + 3 \times 4$ (f) $4 \times (12 - 3)$

2 A and F, B and H, C and G, D and E

3 Expressions A, B, D and E do not need brackets.

4 A and G, B and H, C and E, D and F

5 (a) 4 (b) 12 (c) 7.5 (d) 1
  (e) 4 (f) 5 (g) 3 (h) 4

Section D (p 57)

1 (a) 6.5 (b) 24.3 (c) 8.5

2 (a) 5 (b) 3.9 (c) 44.8
  (d) 29.6 (e) 32 (f) 6.65

Section E (p 57)

1 (a) 5 (b) 36 (c) 39 (d) 40
  (e) 94 (f) 8 (g) 4 (h) 9

2 (a) [9] [×] [4] [x^2]

  (b) [(] [9] [−] [4] [x^2] [)] [÷] [4]

  (c) [(] [4] [+] [9] [x^2] [)] [÷] [4] [x^2]

  (d) [9] [÷] [(] [9] [−] [4] [x^2] [)]

3 (a) 10.8 (b) 42.25 (c) 5
  (d) 1.75 (e) 4.375 (f) 8.56
  (g) 6 (h) 4.5 (i) 1.4
  (j) 25.5 (k) 13 (l) 8.8

*4 $\dfrac{a^2 - b^2}{a - b} = a + b$

Sections F and G (p 58)

1 (a) ⁻14.5 (b) 1.7 (c) 51.894
  (d) 5.7 (e) 28.7 (f) 15.265 625

2 (a) 5.98 (b) ⁻46.44 (c) 461
  (d) 703 (e) ⁻39.25 (f) ⁻7.7
  (g) 5 (h) 4.5 (i) 6
  (j) 78.06 (k) 8.4

*3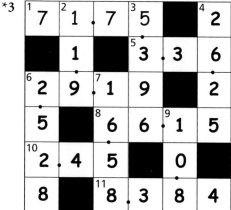

21 Three dimensions

Essential	Optional
Linking cubes Triangular dotty paper	Sheet 170 Collection of three-dimensional shapes, e.g. cartons

Practice booklet pages 59 to 65

Ⓐ **Drawing three-dimensional objects** (p 164)

Linking cubes, triangular dotty paper

◊ If objects which are 'mirror images' of each other are counted as different, there are 8 objects which can be made with four cubes and 29 objects with five cubes.

Four cubes

 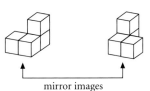

mirror images

Five cubes

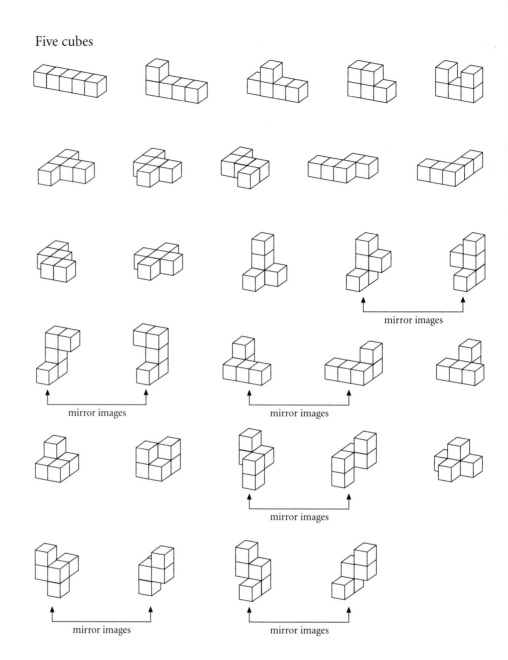

mirror images

mirror images

mirror images

mirror images

mirror images

mirror images

B Views (p 164)

The choice of 'front' and 'side' view is arbitrary.

C Volume of a cuboid (p 166)

Before doing work on volume it is a good idea to ask pupils how they would define 'cuboid'; for example, how would they describe a cuboid over the phone? This is a preparation for the more tricky question later of defining 'prism'.

D Prisms (p 168)

> Optional: collection of three-dimensional shapes (e.g. cartons)

◊ Defining 'prism' is harder than defining
'cuboid'. This is a task which could be set
to pupils working in pairs or groups. It is
useful to have a number of three-
dimensional shapes available to help
clarify later what is and what is not a
prism. Have some counter-examples ready;
for example, the suggested definition 'same
shape all through' can be met with a
twisted pile of paper.

◊ The only prisms used at this stage are 'right' prisms, where a plane shape
(the cross-section) is translated in a direction perpendicular to its plane.

E Volume of a prism (p 168)

F Nets (p 171)

> Optional: sheet 170

Pupils should be encouraged to do as much as they can by visualising the
three-dimensional shapes – they should cut out and fold nets only when
they cannot visualise the shapes, or to check their thinking.

◊ Nets P, Q and R are on the optional resource sheet 170. Having discussed
their sketches, perhaps first in small groups, and then as a whole class,
pupils could make the solids.

◊ P   square-based
      pyramid

Q cube with
  corner
  sliced off

R 'house' or
  cube with
  triangular
  prism

◊ Useful discussion questions are 'What is the same about each solid? What
is different?' It is hoped that pupils will consider and discuss faces,
vertices and edges, but do not force this.

G Surface area (p 173)

B Views (p 164)

B1 (a) 　(b)

B2 (a) Front Side

Top

(b) Front Side

Top

(c) Front Side

Top

B3 (a) A P, B R, C S, D Q

(b) View from T

B4 Mug A, G

Spoon B, D

Toilet roll C, E

Book F, H

or

C Volume of a cuboid (p 166)

C1 (a) 40 cm³　(b) 60 cm³　(c) 24 cm³

C2 (a) 30 cm³　(b) 48 cm³
(c) 28 cm³　(d) 36 cm³

C3 10.5 cm³. You can see nine whole cubes and three halves.

C4 12.25 cm³

C5 (a) 36 cm³　(b) 45 cm³
(c) 16.75 cm³　(d) 20.25 cm³

C6 $a = 4$ cm　　$b = 9.6$ cm　　$c = 1.5$ cm
$d = 4$ cm　　$e = 3.63$ cm (to 2 d.p.)

C7 For example, 10 by 20 by 24 cm,
10 by 16 by 30 cm, 12 by 16 by 25 cm;
volume = 4800 cm³ in each case

E Volume of a prism (p 168)

E1 (a) 60 cm³　(b) 98 cm³　(c) 125 cm³
(d) 40 cm³　(e) 96 cm³　(f) 123 cm³

E2 (a) 48 cm³　(b) 84 cm³　(c) 186 cm³

E3 (a) 60 cm³　(b) 94.5 cm³　(c) 180 cm³
(d) 60 cm³　(e) 108 cm³

E4 2.79 m³

E5 0.55 m = 55 cm

E6 40 cm

E7 4000 cm²

E8 40 m²

E9 The pupil's sketches

E10 The pupil's sketches

E11 The volume is multiplied by
(a) 2　(b) 4　(c) 8

*E12 7 cm by 14 cm by 21 cm
(width, height, length)

*E13 11 cm

F Nets (p 171)

F1 (a) The missing face could go on the net in any one of four positions.

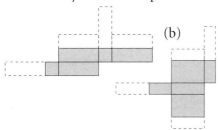
(b)

F2 (a) P and Q are nets of a cube, R is not.

(b) The arrangements which make a cube are

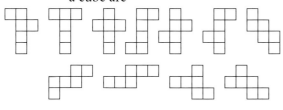

F3 (a) 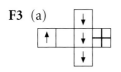   (b)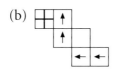

F4 (a)

(b)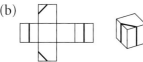

F5 (a) Square based pyramid

(b) Triangular prism

(c) Tetrahedron

(d) Triangular prism ('wedge')

(e) Hexagonal prism

F6 (a) There are several possibilities. Here are two.

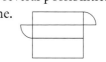

(b) There are several possibilities. Here is one.

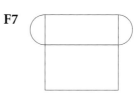

F7

F8

G Surface area (p 173)

G1 126 cm^2

G2 (a) 122 cm^2 (b) 202 cm^2
(c) 134.5 cm^2 (d) 35.5 cm^2

G3 Not necessarily. Pupils should give at least one example of a pair of cuboids with the same volume but different surface areas (e.g. 2 by 4 by 2 [surface area 40] and 8 by 2 by 1 [surface area 52]).

G4 192 cm^2
(The missing edge length is 12 cm.)

G5 248.5 cm^2
(The missing edge length is 6.5 cm.)

G6 9 cm

G7 16 cm^3

What progress have you made? (p 174)

1 The pupil's drawing of a 5-cube object

2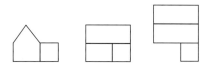

3 30 cm³

4 70 cm³

5 0.42 m (to 2 d.p.)

6 (a) Here is one
        possibility.

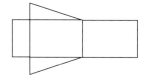

   (b) Here is one
        possibility.

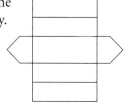

7 67 cm²

Practice booklet

Section B (p 59)

1 (a)   (b)

2 front 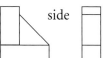 side top

3 A   B C

Section C (p 59)

1 (a) 208 cm³   (b) 675 cm³
   (c) 1700 cm³   (c) 189 cm³

2 $a = 2$ cm, $b = 1.25$ cm, $c = 5$ cm

3 (a) 1 000 000 cm³
   (b) 1000 litres   (b) 72 000 litres

Section E (p 60)

1 (a) 46 cm³   (b) 36 cm³
   (c) 90 cm³   (d) 90 cm³

2 (a) 30 cm³   (b) 42 cm³

3 (a) 120 cm³   (b) 25 cm³

4 (a) 280 cm³   (b) 560 cm³

5 24 cm

6 $a = 2.5$ cm, $b = 9$ cm, $c = 3$ cm

Section F (p 62)

1 (a)   (b)

   (c)   (d)

2 There are two different possible dice, one
   with the numbers 2 3 5 4 going clockwise
   around the number 1,

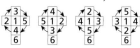

   and the other with the numbers
   2 3 5 4 going anticlockwise around the
   number 1.

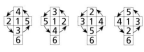

3 B, E, F

116 • *21 Three dimensions*

4 (a)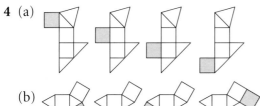

(b)

5 (a)

(b)

(c)

6 (a) Equilateral triangle

(b) Square

(c) Rhombus

(d) Regular hexagon

Section G (p 64)

1 (a) 14 cm (b) 84 cm² (c) 104 cm²

2 (a) $2 \times (4 \times 6) + (7 \times 20) = 188$ cm²

(b) $2 \times (7 \times 2) + (6 \times 18) = 136$ cm²

(c) $2 \times (3 \times 7) + (3 \times 20) = 102$ cm²

3 (a) 28 cm (b) 7 cm

(c) 196 cm² (d) 276 cm²

4 (a) $(2 \times 96) + (5 \times 56) = 472$ cm²

(b) $(2 \times 84) + (8 \times 50) = 568$ cm²

(c) $(2 \times 84) + (6 \times 56) = 504$ cm²

5 (a) (i) 6 cm² (ii) 24 cm² (iii) 54 cm²

(b) (i) 1 cm³ (ii) 8 cm³ (iii) 27 cm³

(c) 96 cm² and 64 cm³

(d) 864 cm² and 1728 cm³

6 (a) (i) 94 cm² (ii) 376 cm²

(b) (i) 70 cm² (ii) 280 cm²

The surface area of the second cube in each pair is 4 times the surface area of the first.

7 (a) (i) 70 cm² (ii) 630 cm²

(b) (i) 40 cm² (ii) 360 cm²

The surface area of the second cube in each pair is 9 times the surface area of the first.

22 Finding formulas

This is about finding a formula for the *n*th term of a sequence. At first the sequences are related to spatial patterns which can be analysed to give an expression for the *n*th pattern. The later part of the unit focuses on linear sequences of numbers.

> **Optional**
> Squared paper
> OHP, matches
>
> **Practice booklet** pages 66 to 70

 A Maori patterns (p 175)

> Optional: squared paper

◊ The objective of the introduction is to bring out the different ways of tackling the task of finding a general rule. There is a sequence of patterns on p 175 which can be used as a basis for this.

You could set pupils (possibly working in pairs) the problem of finding how many crosses there will be in pattern 10, then 100 and to see if they can write a formula for the number in pattern *n*. It is important that everybody should have a go at this task before they hear anyone's answers.

◊ It is essential in the subsequent discussion to bring out the two ways of generalising: geometrical and numerical.

Geometrical – Pupils may have different ways of breaking up the pattern but this is probably the most common:

'Pattern 10 will have 10 crosses to the left, 10 to the right, 10 below and one in the middle, leading to $3n + 1$ for pattern n.'

Numerical – 'Make a table of the number of crosses in order.

Pattern number	1	2	3	4	5
Number of crosses	4	7	10	13	16

The number of crosses goes up by 3 each time, so the formula must start with $3n$. Pattern 1 has 4 crosses, so the formula must be $3n + 1$.'

At this stage, pupils using the numerical approach are unlikely to see how to get from 'goes up in 3s' to '$3n$'. This is covered later in the unit.

In any case, the geometric approach is better in this section and the next because some of the patterns give quadratic sequences.

A3, A4 These questions give opportunities to discuss equivalent expressions. For example in question A3:

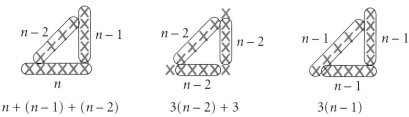

$$n + (n - 1) + (n - 2) \qquad 3(n - 2) + 3 \qquad 3(n - 1)$$

B Matchstick patterns (p 177)

'Using an OHP and matches worked well as a teaching aid.'

Optional:	OHP, matches

C Sequences (p 178)

◊ You could start by asking pupils to find as many different ways as they can of continuing the two sequences given (1, 2, 4, … and 2, 4, 6, …).

Bring out the two different ways of giving a rule for a sequence. For example, the term-to-term rule for 2, 4, 6, … is 'add 2', and the position-to-term rule is 'multiply the position (term number) by 2'.

Of course, when using a term-to-term rule, you have to know the first term as well.

In case pupils are short on ideas for continuing the two sequences, here are some you can mention:

1, 2, 4, 8, 16, 32, ... (multiply by 2)

1, 2, 4, 7, 11, 16, ... (add 1, add 2, add 3, ...)

1, 2, 4, 5, 7, 8, ...   (add 1, add 2, add 1, add 2, ...)

1, 2, 4, 9, 23, 64, ... (multiply by 3 and subtract 1,
         multiply by 3 and subtract 2,
         multiply by 3 and subtract 3, ...)

2, 4, 6, 8, 10, ...   (add 2)

2, 4, 6, 10, 16, 26, ...(add the previous two terms)

◊ Emphasise the fact that although there may be an 'obvious' rule for continuing a sequence, it does not follow that this is the only possible rule. However, pupils should assume that when asked for *the* rule for a sequence, the obvious one is intended.

◊ Explain that a linear sequence is so called because if the terms are plotted on a graph the points are in a straight line. (This point is returned to in a later unit.)

C4 Encourage pupils to think carefully about their methods and to try to avoid trial and improvement. It is a good idea for them to compare methods.

Ⓓ **Sequences from rules** (p 179)

Ⓔ **Finding a formula for a linear sequence** (p 179)

◊ Give everybody a chance to see if they can find a rule for the *n*th term of each of the three sequences given. If pupils are struggling with B or C, a hint along these lines is useful:

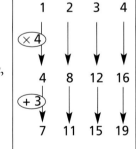

Start with the sequence 1, 2, 3, 4, 5, ... for which the *n*th term is just *n*.

Choose a number to multiply every term by, say 4.

Notice that the new sequence, whose *n*th term is 4*n*, goes up by 4 each time.

Choose a number to add to every term, say 3.

Notice that the new sequence still goes up in 4s, and this is true whatever is added or subtracted.

◊ For the *n*th term of A, some pupils may produce an expression with '+3' in it. They are confusing the relationship across the table with the relationship downwards.

◊ Ask pupils to explain their own approaches to B and C, before focusing on the approach used at the top of page 180.

🅕 **Decreasing linear sequences** (p 181)

This work could be followed up by the activity 'Making a sequence' in *Using a spreadsheet*.

🅐 **Maori patterns** (p 175)

A1 (a)

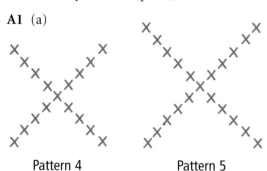

Pattern 4   Pattern 5

(b) Pattern 10 has four 'arms' each of 10 crosses and one cross in the middle.

(c)
Pattern number	1	2	3	4	5
Number of crosses	5	9	13	17	21

(d) (i) 41   (ii) 201   (iii) 401

(e) $4n + 1$

A2 (a)

Pattern 4   Pattern 5

(b) Pattern 10 has two rows of 10 crosses and one cross at each end.

(c) 202

(d) $2n + 2$

(e) Pattern 27

A3 (a)

Pattern 3   Pattern 5

(b) 297

(c) $3(n - 1)$ or equivalent

A4 (a)

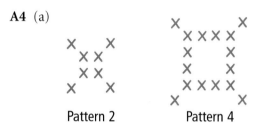

Pattern 2   Pattern 4

(b) Pattern 10 is an open square with 10 crosses on each side with an extra cross at each corner.

(c) 400

(d) $4n$

(e) Pattern 21

A5 (a)

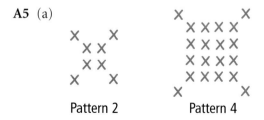

Pattern 2   Pattern 4

(b) Pattern 10 is a 10 by 10 square of crosses with an extra cross at each corner.

(c) 104

(d) $n^2 + 4$

(e) Pattern 15

*A6 $n(n + 1)$ or equivalent

A7 The pupil's design, with formula

$\mathbb{B}$ Matchstick patterns (p 177)

B1 $2n$

B2 $3n + 1$

B3 $5n + 1$

B4 $6n - 2$

B5 $8n + 7$

*B6 $2n(n + 1)$

$\mathbb{C}$ Sequences (p 178)

C1 (a) Add 2, then 3, then 4, … and so on.

(b) 28

C2 (a) (i) Add 5 (ii) 44

(b) (i) Add 1, 3, 5, … (odd numbers)

(ii) 51

(c) (i) Add 2, 4, 8, … (double each time)

(ii) 255

(d) (i) Subtract $\frac{1}{4}$

(ii) $1\frac{1}{4}$

(e) (i) Multiply by 2

(ii) 384

(f) (i) Add the two previous terms

(ii) 47

C3 (a) Linear (b) Non-linear

(c) Non-linear (d) Linear

(e) Linear

C4 Puzzle 1: 10th term is 49
Puzzle 2: 6th term is 17
Puzzle 3: 1st term is 7
Puzzle 4: 8th term is 30
Puzzle 5: 11th term is 49
Puzzle 6: 6th term is 15
Puzzle 7: 7th term is 41

$\mathbb{D}$ Sequences from rules (p 179)

D1 (a) 1, 4, 7, 10, 13

(b) The differences are 3, 3, 3, 3, …
It is a linear sequence.

D2 (a) $4n + 2$ (b) $n + 4$ (c) $5n - 8$

(d) $n^2 - 1$ (e) $8n - 3$

D3 (a) (i) 5, 10, 15, 20, 25, 30

(ii) Linear

(iii) 500

(b) (i) 6, 11, 16, 21, 26, 31

(ii) Linear

(iii) 501

(c) (i) 4, 8, 12, 16, 20, 24

(ii) Linear

(iii) 400

(d) (i) 1, 5, 9, 13, 17, 21

(ii) Linear

(iii) 397

(e) (i) $^-6, ^-3, 0, 3, 6, 9$

(ii) Linear

(iii) 291

(f) (i) 1, 4, 9, 16, 25, 36

(ii) Non-linear

(iii) 10 000

(g) (i) 49, 48, 47, 46, 45, 44

(ii) Linear

(iii) $^-50$

(h) (i) 2, 5, 10, 17, 26, 37

(ii) Non-linear

(iii) 10 001

(i) (i) 28, 26, 24, 22, 20, 18

(ii) Linear

(iii) $^-170$

(j) (i) 12, 6, 4, 3, 2.4, 2

(ii) Non-linear

(iii) 0.12

E Finding a formula for a linear sequence (p 179)

E1 (a) $3n + 7$ (b) $6n + 1$

E2 (a) $6n - 5$ (b) $0.5n + 3.5$
     (c) $5n - 11$ (d) $\frac{1}{4}n + 5$

E3 (a) The pupil's explanation, possibly: 'He is mixing up the formula for the nth term with the rule for going from one term to the next.'
     (b) $2n + 6$

E4 Sequence 1: constant difference = 5
Sequence 2: $6n + 5$
Sequence 3: $2n - 1$
Sequence 4: $\frac{1}{2}n + 6\frac{1}{2}$
Sequence 5: $2n - 5$
Sequence 6: 3rd term could be 10, 40, 90, 160, … (all of the form $10p^2$)

F Decreasing linear sequences (p 181)

F1 (a) $53 - 3n$
     (b) $101 - n$

F2 (a) $100 - 2n$; $^-100$
     (b) $46 - 6n$; $^-554$
     (c) $50.5 - 0.5n$; 0.5
     (d) $7.2 - 0.2n$; $^-12.8$

What progress have you made? (p 181)

1 $2n + 4$

2 (a) 10 (b) $^-2$

3 (a) $6n + 5$ (b) $9n - 7$

4 $23 - 3n$

Practice booklet

Sections A and B (p 66)

1 (a) The pupil's sketch of patterns 4 and 5
     (b) The pupil's explanation of pattern 10

(c)

Pattern	1	2	3	4	5	10	100	n
Tables	4	6	8	10	12	22	202	$2n+2$

     (d) Pattern 7

2 (a) The pupil's sketch of patterns 3 and 5
     (b) The pupil's explanation of patterns 10 and 20

(c)

Pattern	3	4	5	6	10	20
Tables	8	12	16	20	36	76

     (d) $4(n - 1)$ or equivalent
     (e) Pattern 25

3 (a) (i)

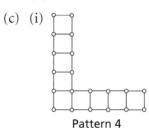

Pattern 4

       (ii) $2n + 1$
     (b) (i)

Pattern 4

       (ii) $3n + 1$
     (c) (i)

Pattern 4

       (ii) $4n + 4$

4 $(n + 1)^2 + n^2$ or equivalent

Sections C and D (p 68)

1 (a) (i) Multiply by 2
       (ii) 1536
     (b) (i) Add the previous two terms
       (ii) 191
     (c) (i) Add 2
       (ii) 112
     (d) (i) Add 1, then 2, then 3, …
       (ii) 55

(e) (i) Subtract 2

    (ii) 7

(f) (i) Multiply by 3 and subtract 1 or add 1, 3, 9, 27 …

    (ii) 9842

2 (a) (i) 3, 6, 9, 12, 15, 18, 21, 24

    (ii) Linear

    (iii) 150

(b) (i) $^-2, ^-1, 0, 1, 2, 3, 4, 5$

    (ii) Linear

    (iii) 47

(c) (i) 9, 8, 7, 6, 5, 4, 3, 2

    (ii) Linear

    (iii) $^-40$

(d) (i) 6, 9, 14, 21, 30, 41, 54, 69

    (ii) Non-linear

    (iii) 2505

(e) (i) 6, 8, 10, 12, 14, 16, 18, 20

    (ii) Linear

    (iii) 104

(f) (i) 42, 21, 14, 10.5, 8.4, 7, 6, 5.25

    (ii) Non-linear

    (iii) 0.84

Section E (p 66)

1 (a) (i) $3n + 4$ (ii) 304

(b) (i) $5n + 2$ (ii) 502

(c) (i) $7n - 4$ (ii) 696

(d) (i) $5n - 5$ (ii) 495

(e) (i) $6n - 7$ (ii) 593

(f) (i) $0.75n + 4.25$

    (ii) 79.25

2 (a) (i) 9, **11**, **13**, 15   (ii) $2n + 7$

    (iii) 57

(b) (i) 2, **10**, **18**, 26   (ii) $8n - 6$

    (iii) 194

(c) (i) $^-1$, **2**, 5, **8**   (ii) $3n - 4$

    (iii) 71

(d) (i) $^-$**2**, 3, **8**, 13   (ii) $5n - 7$

    (iii) 118

3 (a) 11, 15 and 19   (b) $4n + 3$

4 27th term

5 (a) £47.50   (b) £$(40 + 2.5n)$

(c) £77.50   (d) 24 weeks

Section F (p 70)

1 (a) (i) $50 - 2n$ (ii) 10

(b) (i) $21 - n$ (ii) 1

(c) (i) $105 - 5n$ (ii) 5

(d) (i) $10 - 3n$ (ii) $^-50$

2 (a) (i) 6, **4**, **2**, 0   (ii) $8 - 2n$

    (iii) $^-52$

(b) (i) 15, **14**, **13**, 12   (ii) $16 - n$

    (iii) $^-14$

(c) (i) $^-2$, $^-$**5**, $^-8$, $^-$**11**   (ii) $1 - 3n$

    (iii) $^-89$

(d) (i) **42**, 34, **26**, 18   (ii) $50 - 8n$

    (iii) $^-190$

3 (a) (i) $4n - 7$ (ii) 393

(b) (i) $9 - n$ (ii) $^-91$

(c) (i) $4n + 10$ (ii) 410

(d) (i) $15n - 5$ (ii) 1495

(e) (i) $10 - 2n$ (ii) $^-190$

(f) (i) $20 - 0.5n$ (ii) $^-30$

Ratio

Essential	**Optional**
Sheet 190	Sheet 189
Triangular dotty paper is needed for Practice booklet page 74	
Practice booklet pages 71 to 75	

Ⓐ Stronger, darker, sweeter, happier, … (p 182)

◊ The questions are meant to be exploratory, to see how pupils think about problems involving ratio and proportion. The focus is on the arguments they use.

◊ If answers are incorrect, or if pupils have no way of telling, then do not feel you have to teach the correct method here and now. There will be an opportunity to return to the questions later in the unit.

◊ There are several different strategies for each problem. Let pupils explain their methods to each other. Don't hold up one correct method as superior to others.

◊ A common error is to use differences. For example, 2 : 3 might be thought equal to 4 : 5.

◊ The answers are: **1** A; **2** A; **3** B; **4** B; **5** A; **6** the first; **7** A; **8** A; **9** A

B Ratio notation (p 184)

B3(c) If a hint is needed, point out that 3 litres of blue and 2 litres of yellow will make 5 litres of dark green.

C Darker, lighter (p 185)

> Optional:
> The recipes are printed on sheet 189 for cutting up into cards.

◊ The recipes in order from darker to lighter are (reading across):

    M (4:1)   K (3:1)   B, N (4 : 2 = 2:1)

    G (3:2)   P (4:3)   D, J, L, O (1:1 = 2:2 = 3:3 = 4:4)

    C (3:4)   A (2:3)   E, F (2 : 4 = 1:2)   I (1:3)   H (1:4)

◊ To make a recipe between, say, 4:1 and 3:1, some pupils may say 3.5:1. This is correct but you could also ask if they can express it in whole numbers.

D Working with ratios (p 186)

E Sharing in a given ratio (p 188)

F Ratios in patterns (p 189)

> Sheet 190

◊ Pupils need to find a repeating unit from which the pattern is made. F5 is quite tricky!

G Comparing ratios (p 191)

H Writing a ratio as a single number (p 192)

There are many common instances where a ratio is given as a single number. For example the ratio of the circumference of a circle to its diameter is expressed as π rather than as $\pi : 1$.

B Ratio notation (p 184)

B1 (a) 12 (b) 5

B2

Tins of blue	Tins of yellow
8	**20**
14	**35**
10	25
24	60

B3 (a) 9 (b) 10

(c) 30 litres blue, 20 litres yellow

B4 (a) 15 litres blue, 10 litres yellow

(b) 9 litres blue, 3 litres red

(c) 8 litres blue, 4 litres red

D Working with ratios (p 186)

D1 (a) 15 litres (b) 10 litres

D2 (a) 15 litres (b) 4 litres

D3 (a) 1:2 (b) 12 (c) 20

D4 4:1

D5 4:3

D6 (a) 4:3 (b) 3:4

D7 (a) 3:1 (b) 3:2 (c) 2:3 (d) 4:7

D8 **1:5** = 4:20 = 20:100 = 2:10

4:6 = **2:3** = 10:15

3:1 = 9:3 = 12:4

3:2 = 12:8

D9 10 litres blue, 20 litres yellow

D10 5 litres red, 15 litres white

*D11 (a) 15 litres

(b) 5 litres black, 25 litres white

(c) 4 litres black, 10 litres white

(d) 30 litres

(e) 15 litres black, 25 litres white

E Sharing in a given ratio (p 188)

E1 Stuart £8, Shula £4

E2 Dawn £15, Eve £5

E3 Beric £9, Betty £12

E4 (a) £16, £4 (b) £24, £36

(c) £15, £9 (d) £25, £20

(e) £7.50, £5 (f) £4.50, £1.50

(g) £10, £7.50 (h) 80p, £1

E5 (a) James £750, Sarah £1250

(b) 1 year: James £800   Sarah £1200
2 years: James £833+   Sarah £1166+
3 years: James £857+   Sarah £1142+
4 years: James £875   Sarah £1125, etc

(c) James's share goes up as grandmother lives longer.

E6 Alan £50, Bertha £100, Cyril £750

E7 (a) £3, £4.50, £6

(b) £11, £16.50, £38.50

F Ratios in patterns (p 189)

F1 2:1

F2 (a) 2:1 (b) 3:2 (c) 1:1

F3 1:1

F4 (a) 2:1 ▨☐

(b) 3:2 (▨▨☐☐ downwards,
or ▨▨☐▨ across)

(c) 1:3 ▨☐/☐☐

(d) 4:5 ▨☐▨/☐▨☐/▨☐▨

F5 A possible repeating unit is shown in each case.

(a) 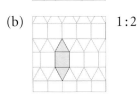 1:1

(b) 1:2

(c)  1:2

(d) 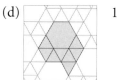 1:8

(e)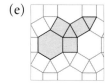
(i) 1:3
(ii) 1:2
(iii) 2:3

G **Comparing ratios** (p 191)

G1 Royal

G2 (a) 3:5 = **15**:25 (b) 7:8 = **28**:32
(c) 11:15 = **66**:90 (d) 4:9 = 36:**81**

G3 Squirrel grey   black : white = 39:15
Thundercloud grey black : white = 40:15
Thundercloud grey is darker.

G4 Jasmine's   juice : water = 55:88
Nita's   juice : water = 56:88
Nita's is stronger (just!).

G5 Initial ratio
        juice : water = 14:9 = 154:99
Final ratio
        juice : water = 17:11 = 153:99
The final drink was weaker (just!).

Challenge
        Red : green = 3:5 = 12:20
        Green : blue = 4:7 = 20:35
        So red : blue = 12:35

H **Writing a ratio as a number** (p 192)

H1 (a) 2   (b) 1.6 (c) 1.5 (d) 0.75

H2 The window is square.

H3 Hamster 0.2   Cat 0.1   Elephant 0.004

H4 A 0.44...   B 0.4...   A is steeper.

What progress have you made? (p 193)

1 (a) 20 litres   (b) 6 litres

2 (a) 2:5   (b) 2:3

3 Gavin £16, Susan £20

4 Peter £12.50, Paul £37.50, Mary £50

5 A black : white = 25:40
  B black : white = 24:40
  A is darker.

6 0.16

Practice booklet

Section B (p 71)

1 (a) (i) 15g   (ii) 36g
  (b) (i) 10g   (ii) 14g

2

Pure gold (g)	Other metals (g)
6	**10**
12	**20**
15	**25**
9	15
18	30
45	75

3 (a) 6g silver, 3g copper
  (b) 10g copper, 330g gold
  (c) 495g gold, 15g copper

Section D (p 72)

1 (a) 4:3   (b) 3:4

2 (a) 2:3   (b) 3:1   (c) 3:5
  (d) 5:2   (e) 7:6

3 (a) 2:3   **6**:9   10:**15**
  (b) 4:1   **12**:3   8:**2**
  (c) 6:2   **3**:1   18:**6**
  (d) 4:10   2:5   **6**:15
  (e) 30:40 **6**:8   3:**4**
  (f) 7:3   21:**9**   **28**:12

4 20:30 and 6:9
    21:6 and 7:2
    9:3 and 12:4
    3:5 and 12:20
    12:9 and 8:6
    8:20 and 2:5

5 (a) 9   (b) 4
    (c) 15 loam and 5 sand

6 (a) 4   (b) 12
    (c) 15 blue and 10 red

Section E (p 73)

1 (a) £4 and £6   (b) £10 and £15
    (c) £32 and £48   (d) £17 and £25.50

2 (a) £10   (b) £12
    (c) £18.75 and £11.25

3 (a) 20 litres blue, 16 litres green
    (b) 7.5 litres blue, 6 litres green
    (c) 9.5 litres blue, 7.6 litres green

4 (a) £20 £40 £60 (b) £48 £24 £48
    (c) £24 £36 £60 (d) £20 £30 £70

5 (a) £0.50 £1 £2
    (b) £0.70 £1.05 £1.75
    (c) £0.75 £1.25 £1.50
    (d) £1 £1.20 £1.30

Section F (p 74)

1 (a) 3:4   (b) 2:3
    (c) 3:1   (d) 3:5

2 One way to do this is to take a row of 10 triangles and colour 6 of them and leave 4 blank. If this block is repeated, the pattern will have the desired ratio of grey to white.

Section G (p 75)

1 A sheets : envelopes = 40:15 = 120:45
    B sheets : envelopes = 25:9 = 125:45
    B has the higher ratio.

2 Rough puff, Flan, Choux, Flaky, Puff

Section H (p 75)

1 (a) 8 cm by 13 cm
        13 cm by 21 cm
        21 cm by 34 cm
    (b) A 2
        B 1.5
        C 1.66…
        D 1.6
        The next three rectangles give
        1.625 1.615… 1.619…
    (c) The answers alternate smaller and bigger. They get closer to 1.618 033… , the golden ratio.

24 Using a spreadsheet

A number of activities relating to different topic areas are gathered together here. If your class has easy access to computers, then you could dip in whenever appropriate. If, however, you have access only at specific times, you could use the unit as a programme of work for your computer sessions. Alternatively, you could ask for the activities to be used in IT lessons which focus on the use of a spreadsheet.

> **Essential**
> Spreadsheet, or similar graphic calculator facility

Spot the formula (p 194)

◊ Some pupils may invent formulas that are impossibly difficult to spot.

◊ Opportunities to discuss equivalent formulas may arise, for example = (A1 + 3)*2 and = 2*A1 + 6.

◊ It is interesting to discuss strategies. For example, putting 100 or 1000 into a formula can often tell you a lot.

Making a sequence (p 195)

◊ You may need to introduce or revise the process of filling down a formula (or 'drag and drop').

◊ Pupils may find a formula for going from one term to the next. Although this is valid, get them to focus on finding a formula that works **across**, calculating the terms of the sequence from the numbers 1, 2, 3, …

Big, bigger, biggest (p 195)

The idea of using decimals may not occur at first.
Here are some solutions (others are possible in some cases):

1 (a) 2, 2, 3, 7   (b) 2, 3, 3, 6   (c) 3.5, 3.5, 3.5, 3.5 give 150.0625

2 (a) 3, 3, 4, 5 or 2.5, 4, 4, 4.5   (b) 2, 2.5, 5, 5.5

   (c) 3.75, 3.75, 3.75, 3.75 give 197.753 906 25

Ways and means (p 196)

Pupils will soon see that the sequence tends to a limit. The challenge is to find how the limit is related to the starting numbers. For two starting numbers a, b the limit is $\frac{1}{2}(a + 2b)$, for three starting numbers a, b, c it is $\frac{1}{6}(a + 2b + 3c)$, and for four, a, b, c, d, it is $\frac{1}{10}(a + 2b + 3c + 4d)$.

Parcel volume (p 196)

The main focus here is on working systematically. As with many problems of this type it helps to restrict the number of variables. If the spreadsheet is set up in the most obvious way (as shown in the pupil's book), it may be difficult to keep track of the restrictions on the variables.

Ask pupils how they could build in the connection between length and girth (which must add up to 3 m for maximum volume). One way is to use a formula equivalent to '3 – girth' in the 'length' column. As girth is already defined as 2(height + width), this reduces the variables to two: height and width.

The maximum volume of $0.25 \, m^3$ occurs when the width and height are both 0.5 m and the length is 1 m.

Furry festivals (p 197)

Pupils may themselves suggest reducing the number of variables by setting the number of pens equal to 60 minus the total number of T-shirts and badges.

The solution is:

    5 T-shirts, 35 badges and 20 pens

or 6 T-shirts, 14 badges and 40 pens

Breakfast time (p 197)

The essential step is to set up a formula for working out a given percentage of a quantity. Pupils may know that, for example, 8 in the percentage column has to be converted to 0.08 but may not immediately realise that the operation needed to do this is ÷ 100.

Mean, median, range (p 198)

In this activity pupils investigate the effect on the mean, median and range of changes in individual data items or transformations of all the data items.

Solution to question 4:

(a) You could increase either 2, 4, 6, 13, 15 or 17 by 1.

(b) Change 13 to 24.

(c) Change 11 to 2, or 13 to 4, or 15 to 6.

(d) Change 10 to 22.

Functions and graphs

Essential	Optional
2 mm graph paper	OHP

Practice booklet pages 76 to 77 (graph paper needed)

Ⓐ From table to graph (p 199)

2 mm graph paper

◊ The main points to be covered in the introduction are:
  • We can make a table from a rule given in words.
  • The values in the table can be plotted as points.
  • When the variables are continuous, a line can be drawn through the points to show the relationship between the variables.

Ⓑ Functions (p 201)

Ⓒ Spot the function (p 202)

Optional: OHP

This is similar to the activity 'Spot the rule' in *Book 1*. You will need a large grid on the board or OHP with x and y axes.

◊ You can start the activity by telling the class that you are thinking of a rule linking x and y, for example $y = x + 2$. Ask a member of the class to give you a value of x; work out y and plot the corresponding point on the grid. Continue until someone can tell you your rule.

Then ask a pupil to take over with a rule of their own.

◊ If pupils are adventurous and use rules involving, for example, squaring, then you can ask what kinds of rule give points which lie in a straight line. (If nobody thinks of squaring, then do so yourself to bring out the distinction between linear and non-linear functions and to explain why the word 'linear' is used for rules of this kind.)

◊ Again, if nobody else does so, use a rule like $x + y = 10$, and bring out the fact that this can also be written $y = 10 - x$ or as $y = {}^-x + 10$.

◊ Ask pupils if they see any similarities and differences between earlier work on sequences and this work. They should be able to appreciate that

- in the case of a sequence, *n* is restricted to being a whole number (discrete), whereas *x* is continuous
- the values of *y* for *x* = 1, 2, 3, 4, … form a sequence (which helps when pupils are trying to find the equation of a linear graph)

C7 You should point out that in real life relationships are rarely precisely linear (except in cases where linearity is built in, as for example with currency conversion graphs).

Ⓐ **From table to graph** (p 199)

A1 (a)

Gas in tank (kg)	Hours away from base
3	5
4	7
5	9
6	11
7	13
8	15
9	17

(b) The pupil's graph of the table, points joined with a line

(c) 8 hours

(d) $5\frac{1}{2}$ kg

(e) 199 hours

A2 (a)

t	0	1	2	3	4	5	6
h	20	50	80	110	140	170	200

(b) $h = 30t + 20$

(c) The pupil's graph of points from the table

(d) $t = 2.7$ (roughly)

A3 (a)

w	0	2	4	6	8	10
d	100	90	80	70	60	50

(b) The pupil's graph of points from the table, labelled '$d = 100 - 5w$'

(c) $w = 4.6$

(d) 7 tonnes

Ⓑ **Functions** (p 201)

B1 (a)

x	-6	-4	-2	0	2	4	6
y	-5	-3	-1	1	3	5	7

(b) The pupil's graph of $y = x + 1$, labelled

B2 (a)

x	-1	0	1	2	3	4
y	-5	-3	-1	1	3	5

(b) The pupil's graph of $y = 2x - 3$, labelled

B3 (a)

x	-4	-2	0	2	4	6
y	6	4	2	0	-2	-4

(b) The pupil's graph of $y = 2 - x$, labelled

B4 (a) The pupil's table for $y = 5 - 2x$

(b) The pupil's labelled graph of $y = 5 - 2x$

Ⓒ **Spot the function** (p 202)

C1

x	0	1	2	3	4
y	2	5	8	11	14

$y = 3x + 2$

C2

x	0	1	2	3	4
y	10	9	8	7	6

Paula

C3 $y = x + 4$

C4 Prakesh

C5 $y = x - 10$

C6 (a) $y = x + 2$ (b) $y = x - 4$

    (c) $y = 8$

    (d) $y = 6 - x$ or $x + y = 6$

C7 $L = 5W + 20$

C8 $d = 350 - 50t$

*C9 (a) $y = 3x - 5$ (b) $y = 10 - 2x$

    (c) $y = \frac{1}{2}x + 2\frac{1}{2}$

*C10 (a)

C	−15	−10	−5	0	5	10	15	20
F	5	14	23	32	41	50	59	68

(b) and (c)

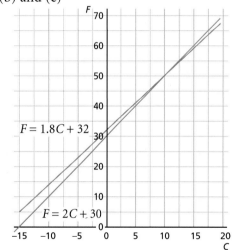

$F = 1.8C + 32$

$F = 2C + 30$

(d) $C = 10$

(e) −15 to 35 (35 is as far above 10 as −15 is below)

What progress have you made? (p 205)

1 The pupil's graph of $y = 2x + 3$

2 (a) $y = 2x + 2$

    (b) $y = 7 - x$ or $x + y = 7$

Practice booklet

Section A (p 76)

1 (a)

t	0	1	2	3	4	5
d	80	100	**120**	**140**	**160**	**180**

    (b) $d = 80 + 20t$

    (c) The pupil's graph from the table above

    (d) $t = 2.4$

2 (a)

t	0	1	2	3	4	5
d	30	**26**	**22**	**18**	**14**	**10**

    (b) The pupil's graph from the table above

    (c) 3.5 minutes

Sections B and C (p 77)

1 (a)

x	−4	−3	−2	−1	0	1	2	3	4
y	**−12**	**−10**	**−8**	**−6**	**−4**	**−2**	**0**	**2**	**4**

    (b) The pupil's labelled graph of $y = 2x - 4$

2 (a)

x	−4	−3	−2	−1	0	1	2	3	4
y	**17**	**15**	**13**	**11**	**9**	**7**	**5**	**3**	**1**

    (b) The pupil's labelled graph of $y = 9 - 2x$

3 (a) $y = x + 7$

    (b) $x + y = 2$ or $y = 2 - x$

    (c) $y = 3x - 3$

4 $h = 35 - 5w$

Review 3 (p 206)

1 (a) 1.215 44   (b) 1.555
  (c) 7.046 718 75

2 (a) 87.75 cm³   (b) 144 cm²

3 (a) (i) $6n - 3$ (ii) $50 - 4n$
  (b) (i) 117   (ii) $^-30$

4 (a) $10 + 6a$ (b) $24b^2$ (c) $^-2c - 5$
  (d) Cannot be simplified
  (e) 3   (f) $14f^2$

5 Jane: 65.2% (to the nearest 0.1%)
  Sinead: 63.3% (to the nearest 0.1%)
  Jane did better.

6 8

7 (a)

t	0	1	2	3	4	5	6
d	25	23	**21**	**19**	**17**	**15**	**13**

  (b) The pupil's straight line graph
  (c) $d = 25 - 2t$ (d) $12\frac{1}{2}$ days

8 (a) 2:3 (b) 2:3 (c) 2:3
  (d) 81, 121.5

9 (a) $n = 9$ (b) $n = ^-7$ (c) $n = 140$

10 (a) $y = x - 4$ (b) $y = 2x + 4$
  (c) $x + y = 4$

11 (a) 5:4   (b) 5:11

12 $\frac{1}{4}$ 0.3 $\frac{8}{25}$ $\frac{17}{50}$ $\frac{7}{20}$ 0.37 $\frac{39}{100}$ $\frac{2}{5}$

13 (a) $x = 32°$ (b) $y = 124°$ (c) $z = 95°$
  with pupil's reasons

14 (a) The totals are equal.
  (b) The pupil's addition crosses
  (c) Sum of red squares
      $= a + (b + c) = a + b + c$
      Sum of blue squares
      $= b + (a + c) = a + b + c$
  (d) The products of the 'red' and 'blue'
      squares are equal.
      Both are equal to $a \times b \times c$

15 $1 \times 1 \times 60$
    $1 \times 2 \times 30$
    $1 \times 3 \times 20$
    $1 \times 4 \times 15$
    $1 \times 5 \times 12$
    $1 \times 6 \times 10$
    $2 \times 2 \times 15$
    $2 \times 3 \times 10$
    $2 \times 5 \times 6$
    $3 \times 4 \times 5$

16 Area of garden = 16^2 m² = 256 m²
    Area of lawn = 256 ÷ 2 = 128 m²
    Length of each side of lawn
    $= \sqrt{128} = 11.31\ldots$ m
    Width of paths
    $= \frac{1}{2}(16 - 11.31\ldots) = 2.34$ m (to 2 d.p.)

Mixed questions 3 (practice booklet p 78)

1 (a) 528 m²   (b) 63 360 m³
  (c) 768 lorry loads

2 (a) 14 cm², 30 cm² (b) $4n + 2$
  (c) 37 cubes   (d) $16n + 6$

3 (a) Ann £7.50, Brian £10, Charlie £12.50
  (b) Ann £8, Brian £10, Charlie £12

4 (a) $3n - 1$ (b) $4n - 11$
  (c) $28 - 2n$ (d) $14 - 9n$

5 (a) $y = x + 5$ (b) $y = x - 4$
  (c) $y = 3 - x$ (d) $y = 2$
  (e) $y = 2x + 3$

6 (a) (i) 105   (ii) $^-95$
  (b) (i) 96   (ii) $^-104$
  (c) (i) $^-97$ (ii) 103
  (d) (i) 2   (ii) 2
  (e) (i) 203   (ii) $^-197$

7 (a) 2.76   (b) 3.85

8 25%

9 (a) 1 person   (b) 2 people

    (c) 2.48 people

10 (a) ⁻32   (b) $-32 + 8(n-1)$

11 (a) RL = red left   RR = red right
       BL = blue left   BR = blue right
       GL = green left   GR = green right

⊗ = L and R choices

2nd glove / 1st glove: RL RR BL BR GL GR

    (b) $\frac{9}{15} = \frac{3}{5}$ (c) $\frac{3}{15} = \frac{1}{5}$

*__12__ 9 : 31